WRITING THE GOODLIFE

PRISCILLA SOLIS YBARRA

WRITING THE GOODLIFE

Mexican American Literature and the Environment

THE UNIVERSITY OF ARIZONA PRESS
TUCSON

The University of Arizona Press
www.uapress.arizona.edu

Printed in the United States of America
21 20 19 18 17 16 6 5 4 3 2 1

ISBN-13: 978-0-8165-3200-1 (paper)

Cover design by Nicole Hayward

Publication of this book is supported in part by a grant from the Association for the Study of Literature and Environment (ASLE), and by the proceeds of a permanent endowment created with the assistance of a Challenge Grant from the National Endowment for the Humanities, a federal agency.

Library of Congress Cataloging-in-Publication Data
Names: Ybarra, Priscilla Solis, author.
Title: Writing the goodlife : Mexican American literature and the environment / Priscilla Solis Ybarra.
Description: Tucson : The University of Arizona Press, 2016. | Includes bibliographical references and index.
Identifiers: LCCN 2015033625 | ISBN 9780816532001 (pbk. : alk. paper)
Subjects: LCSH: American literature—Mexican American authors—History and criticism. | Environmentalism in literature.
Classification: LCC PS153.M4 Y38 2016 | DDC 810.9/86872—dc23 LC record available at http://lccn.loc.gov/2015033625

∞ This paper meets the requirements of ANSI/NISO Z39.48-1992 (Permanence of Paper).

For my parents,
María Higinia Solis Ybarra
and
Encarnación Peña Ybarra

"Sometimes I believe the best I've done as a writer is just to want. Just to keep wanting stories that bring us to our knees, return us to God, humble us in full recognition of our dependence on one another and a flourishing planet."

CHERRÍE MORAGA, *A XICANA CODEX OF CHANGING CONSCIOUSNESS*

"It is from the body, not the mind, that questions arise and answers are explored."

WALTER D. MIGNOLO, *THE DARKER SIDE OF WESTERN MODERNITY*

CONTENTS

ILLUSTRATIONS

FIGURE 1. Dad, Mom, Saulo, and me at Tía Manuela's house in Oak Cliff (Dallas, TX), 1984. Photo from author's personal collection.

PREFACE

The Making of a Chicana Ecocritic

POPULAR IMAGINATION associates environmental writers and environmental literary critics with a life spent hiking mountain trails, kayaking through rapids, or living off the grid. Yet I was born in the ninth most populous area in the United States (Dallas, Texas), and I attended graduate school first in Los Angeles and then in Houston during the late 1990s and early 2000s, when the two competed for the title of city with the most polluted air. My wilderness experience is best described as slim to none, and the first time the environmental movement caught my attention was in 1990 during the twentieth anniversary revival of Earth Day. I was seventeen years old at the time, and Earth Day seemed like a good idea to me, but I couldn't imagine how I might fit into the movement. My family moved to a little town in north Texas when I was four years old, but I never equated my rural life with any kind of environmental experience.

For one, my parents never took us tent camping or hiking when my brother and I were growing up. My father's childhood experience as a Mexican American migrant farmworker, traveling the seasons from Texas to the Midwest and back during the Great Depression, living in tents from necessity, taught him to see camping and hiking as hardships. The only "hiking" he ever saw was of the hitchhiking variety, by need rather than by choice. His graduation from day laborer with no job security to a solid blue-collar career as a truck driver for Affiliated Food Stores gave our family a decidedly motor-oriented lifestyle. My mother regards rural agricultural areas as among the most beautiful possible—she was raised in a farming community outside of Matamoros in the state of Tamaulipas, situated on the northern Mexico border—but the pleasures of a primitive campsite and isolated hiking trails sound dangerous to her. So I was in my twenties by the time I slept in a tent by my own initiative, and even that was just car camping. These days I take part in urban and suburban hikes much more often than the rural type of wanderings, and my few nonurban walks take place on Texas prairies more than in scenic mountain landscapes.

Likewise, the authors whose work I discuss in this book share a daily experience with the natural environment that is urban as well as rural, local at the same time as it is impacted by far-flung migrations. Some are set in densely populated areas and others in remote locales, and, moreover, their works are more apt to portray everyday labor contexts than "wild" or leisure-oriented nature. As a result, these literary works offer little-heard insights about a daily and intimate relationship with the natural environment. One might even wonder what makes these writings environmental in the first place, as they do not fit the mold of "nature writing." I am writing this book because my parents taught me how Mexican Americans have a unique relationship with the land, but I did not see that lesson represented in any of my courses or the scholarship that I read after I left home. My parents sent me to the university to learn what I needed to know in order to succeed. As it turns out, they had already taught me how to make a unique contribution to the field of environmental studies and Chicana/o literary studies.

I know that I am not the only low-income, first-generation college student from an ethnic minority background who aspires to tell her parents' story through her life's work. Explicit or not, their thoughts, stories, and experiences show up on every page of my writing, so you should know that neither of my parents ever attended school. My father, Encarnación Peña Ybarra, was born in 1918 under a tree in Kaufman, Texas, with the growing city of Dallas nearby. At the time, Dallas's finance and banking empires were just getting their start, Love Field was a mere training area for aspiring pilots, and Southern Methodist University had only been around for three years. Busy meeting their basic needs, the Ybarra clan did not take part in the finance, aviation, or educational aspects of the city's enterprise. My father was an older sibling in a group of twelve who, with their parents, traveled the migrant farmworker circuit from near the border with Mexico to near that with Canada and back. He told me stories from his tough childhood and youth during the Great Depression. As

Américo Paredes (born in 1915) puts it in his novel *George Washington Gómez*, the Mexicans couldn't take a job if there were Anglos who wanted it (Paredes 195). If a Mexican did get a job, he was paid, at best, half the wage of an Anglo because the bosses reasoned that Mexicans could get by with less. The bosses tell the Mexicans that the Anglo "needs more money to live on. You can do with less" (Paredes 200). I first heard that story from my dad, and when I read it in Paredes's novel, my dad's voice rang in my ear once again. My father's parents, on the move and always looking for work, figured it wasn't worthwhile to put the children into a school from which they must promptly remove them to look for work farther down the line. Tomás Rivera describes this migrant life in *And the Earth Did Not Devour Him*. Unlike the little boy in that book, who did go to school, my dad helped his family survive by working in the fields, and somehow, eventually, taught himself how to read and write in Spanish and then in English. He even learned to play a pretty good accordion and *bajo sexto* (six-string bass guitar), not to mention the guitar, and put together his own conjunto (band) in Dallas during the 1960s for a while. By the time I showed up in his life, he was a truck driver running a delivery route from Dallas to Houston and Galveston and back. I traveled back and forth on the same road—Interstate 45—when I was going to graduate school at Rice University. I thought of him often, wishing I could tell him just one or two of the things I was learning about, things that in many ways he had already taught me.

My dad died in 1991, just a month short of his seventy-third birthday. I was seventeen years old and halfway through my junior year in high school. Before he died, he tried to catch my adolescent attention long enough to get me to help with the garden he kept in our back yard, and I wish I had paid more attention to what he shared with me about how to help things grow. Still, some lessons took root. I remember one lesson in particular: he would ask my older brother and then me what they taught us in school about Texas history. My father read a lot of books, especially history books, and he wanted us to know the Mexican side, as well as the American and Texan side, of Texas history, even taking us on trips to the Alamo and the San Jacinto Monument. Taking a "busman's holiday," he would drive us to South Texas, hauling our secondhand pop-up trailer behind our yellow-and-white Chevy pickup, the four of us crammed onto the bench seat and our dog Fido (a mostly German Shepherd mutt with floppy ears) lying on the floorboard at our feet. We'd find an RV park to perch our trailer, and my mom would cook our rice and beans, *papas con carne* (a meat and potato dish), and fresh flour tortillas just like at home. My dad took us to the monument, told us to pay attention to him rather than to the guides, and listened to our questions and reactions as if we really had something to say. I recall these trips often and wish I'd had my dad by my side when I first read *Caballero*, Jovita González's novel that tells the Mexican side of the Texas War of Independence from Mexico, and *With His Pistol in His Hand*, Américo Paredes's demystification of the Texas Rangers. My father was the first teacher to show me how to question master narratives.

My mother, María Higinia Solis Ybarra, was born in 1938 on a rural *ejido* (post-Revolution agricultural community) in northern Mexico outside the big border city of Matamoros. She was abandoned by her mother shortly after her birth, and her father was killed when she was only one year old. She was raised with her only sibling, a brother four years older, in her ejido community, called La Venada, by her father's aunt. A pair of orphans in an agricultural community considered themselves lucky to be able to earn their keep by working, and my mother and uncle worked steadily on the lands, some of those lands willed to them by their dead father. Their work kept them from the small local school, but a teacher took personal responsibility for their education and visited them at home when they could learn the early stages of reading, writing, and basic arithmetic. Still, their primary concern was with farming and the accompanying challenges of a rural life. I grew up hearing about how to grow cotton and watermelons and plants that can cure a wide range of ailments. I'm still trying to learn these lessons from my mother, who mastered not only farming techniques but the natural history of the Rio Grande Valley and a boundless knowledge of herbal remedies—all without formal schooling. When I was a kid, I always wondered why everybody else in our little town had neatly trimmed lawns while ours teemed with a chaos of weeds. These days we call those weeds "native species" and we call that style of lawn "native landscaping," and having one designed costs a lot of money. Despite her love for her childhood agricultural community, my mother left her home at the age of eighteen and arrived, young and completely alone, in Houston, Texas, in search of her mother. Arriving in the United States as a young woman with no education, no English, and very limited resources, my mother made her own way, in defiance of the intersectional discriminations she encountered every day. My mother taught me the fierce independence of a strong woman who never thinks twice about taking meaningful risks, forging an unconventional path, and connecting with place through knowledge and cultivation of local ecology. It is her voice I hear when Rudolfo Anaya's Ultima speaks and her remedies that resonate with the *curanderismo* (folk medicine) I find in Patricia Preciado Martin's *Songs My Mother Sang to Me*.

I am extremely fortunate to have my parents' lives to remind me that the ways to address our current environmental (and many other) problems do not always lie in the future or in the new. We can learn so much by paying attention to those around us, even if—or especially because—they are not the ones we conventionally look to for wisdom and insight.

ACKNOWLEDGMENTS

THE SEEDS OF THIS PROJECT first germinated during my undergraduate years at the University of North Texas (UNT), when I was discovering both environmental studies and Chicana/o literature. I never expected to be back at the same university, now as a faculty member, as I finished this project. But I really consider this to be more of a beginning than a completion. So much good work remains to document Chicana/o writing and the environment, and I look forward to continuing to pursue this passion. I take full responsibility for any shortcomings in this book. I cannot, however, take credit for shaping this work on my own. Many relationships have brought this into fruition, both directly and indirectly, and I would like to acknowledge them.

José Aranda gave me unequivocal, but not uncritical, support from the day we first met, and he continues to challenge my thinking to this day. Both his exemplary scholarship and the generous amount of time he spent in thoughtful conversation with me directly shape this project. I only hope that I can begin to approximate his grace and generosity as a mentor and friend. Krista Comer consistently offered sharp criticism, and I am thankful for that, as well as for her friendship. Walter Isle willingly gave his time to my work even as he served as an administrator. I am grateful for that, as well as for his mentorship in the world of ecocriticism. Alex Lichtenstein graciously agreed to join my dissertation committee when I asked him at the eleventh hour, and I am thankful for his helpful commentary from the history side of my literary history. Hamid Naficy was also a wonderful mentor at Rice University.

I worked on these ideas at five different institutions, but Rice University and its Department of English was where these ideas took their initial root. I am grateful for its generosity and flexibility to accommodate an interdisciplinary study, as well as for the financial support to present my work at national and international conferences. Also, I cannot leave unrecognized the unofficial "institution" of *café con plática* and the vibrant intellectual community I found in Houston—Lourdes Alberto, Cecilia Ballí, John Escobedo, Charli Valdez, and Liz Fenton, in particular. Also, I could not have wished for a better cohort at Rice than Ayşe Celikkol. Nuestra Palabra—especially Tony Diaz, Liana Lopez, and Bryan Parras—was also a vital part of my formative years in Houston.

I am fortunate to be part of an extended network of scholars and friends that has nurtured me on this path. The University of California, Los Angeles, was a great place to start, and Theresa Delgadillo, Rafael Pérez-Torres, and Sonia Saldívar-Hull helped me find my way. Stephen Pitti and Alicia Schmidt-Camacho offered me a home base at Yale's American Studies Program during my year in New Haven. I am grateful to them, as I am to my students in the Chicano Culture and Politics class I taught in fall 2004. And of course the rich conversations over sumptuous dinners at Mechanic Street were key ingredients in New Haven: thank you, Donia Mounsef and Natalie Meisner.

My first job at Texas Tech University was a wonderful experience. I was lucky to have such generous and talented mentors and colleagues. I couldn't have made it through without all the love and absolute brilliance of my friends in West Texas. Thank you to my Lubbock family: Sara Spurgeon, Jennifer Snead, Karen Clark, Lynne Fallwell, Carmen Pereira, John Beusterien, Idoia Elola, Curtis Bauer, Kurt Caswell, Kanika Batra, Rich Rice, Susan Tomlinson, Jorge Ramirez, Anna Christina Ribeiro, and Yuan Shu. Juan Muñoz and Jobi Martinez also made me feel at home and gave me much support. Barry Lopez, I really miss crossing paths with you in Lubbock. Thank you also to my students at Texas Tech, with whom I learned so much.

The professionals at the New Mexico State Archives answered questions before I formulated them and anticipated my venues of investigation. Many thanks to them, and to Estevan Rael-Gálvez, for his mentorship and for suggesting I apply for the Scholar's Program.

I have met so many colleagues at conferences and through professional organizations over the years who have become friends and mentors. Thank you especially to Jorge Marcone, Joni Adamson, Scott Slovic, Cheryll Glotfelty, John González, Maria Herrera Sobek, Mary Pat Brady, Sarah Wald, Sarah Jacquette Ray, Estevan Arellano, David Vasquez, Kathleen Dean Moore, Tom Lynch, Stacy Alaimo, Stephanie LeMenager, Annie Merrill, and Cate Sandilands. Thanks also to the readers of this manuscript for your generosity and your advice. Kristen Buckles has been fantastic to work with at the University of Arizona Press. Thank you for your enthusiasm for this project from the start.

The writing workshop, Avanzamos: El Taller Chicana/o, which was the brain storm of José Aranda and which we've led together over the past five years, has been helpful beyond words. Thank you to all the participants who read my work and who brought your brilliant work to share there. I've especially treasured working with the "next generation" through Taller: Amanda Branker Ellis, Lorena Gauthereau, Elena Valdez, Brittany Henry, and Scott Pett. Melina Vizcaíno-Aleman, I'm so happy we met through El Taller, and I look forward to more future collaborations.

UNT welcomed me back when I least expected it, and it has been a great homecoming in so many ways. I'll tell anyone who cares to listen that UNT students bring joy to my teaching in multiple ways. Thanks to all of you who have shared a semester with me along the way. I would especially like to thank graduate students Ashley Reis, Samuel Coronado, and Maricruz Gómez. I look forward to being colleagues with you over the coming years. My new colleagues quickly became my friends, too, and I am so grateful to share this path with you all, especially the fabulous women of La Colectiva, including Mariela Núñez-Janes, Alicia ReCruz, Valerie Martinez, Yolanda Flores Niemann, and Bertina Combes. I must mention a few more cherished UNT colleagues by name, too: Marianne Bueno, Darrell Wanzer Serrano, Kelly Wisecup, Walton Muyumba, Irene Klaver, Brian O'Connor, Brad Leali, Chad Pearson, Sandra Mendiola Garcia, Annette Lawrence, Shay Youngblood, Paula Gaetano-Adi, Alejandro Borsani, Mike Wise, Jennifer Wallach, Barbara Rodman, Deborah Needleman Armintor, and Marshall Armintor. To Laila Amine and Michael Thompson: so many drafts, so many meals, so much laughter; thank you from the bottom of my heart for your grace, generosity, and companionship. You are brilliant colleagues and the best of friends.

It seems strange to thank a place, but I wrote significant portions of this book at Banter, a sadly now-defunct but much-loved coffee shop in Denton, Texas. Thank you, Kat and Ellen, for curating such a nurturing space. It was my second home in Denton, where you welcomed me so warmly, fed me so well, and played the perfect tunes to accompany my marathon writing sessions.

I have been fortunate to have a very understanding network of friends. They put up with my unpredictable fluctuations—when I disappear into my work, when I need to take a break, and when I need to talk about my ideas ad infinitum. Thank you, in particular Rosario Daza (*y* Mateo *y* Chris, *por supuesto*), Sarah Oglesby, and Willem Versluis. I'm getting on a plane to Australia as soon as I possibly can.

J. Baird Callicott saw me through these many long years of writing with true grit and characteristic steadiness. *Mil gracias*, Baird, for your companionship and support in every way. You've read every word of this book many times over, and I thank you for all that and more. We also shared life with the most wonderful dalmatian in the world, who slept soundly and in a perfect ball of black and white fur at my feet during countless hours of this writing. I hope there's lots of good leftovers to eat and endless beaches to romp on wherever you are now, beautiful Whisper. The indomitable Esti still protects me fiercely, showing me love, and taking me for necessary walks everyday.

My ever-patient and understanding family never doubted my aspirations and continues to support my academic migrations. For so many reasons, this work would not be possible without them. *Gracias a mis hermanos y mis hermanas*: Saulo, Manuel, Rick, and Rey; Blanca, Martha, and Rebecca; and to my nieces and nephews, too.

This book is dedicated to my parents, whose stories will always shape my world, inform my work, and fill my heart: my mother, María Higinia Solis Ybarra, and my father, Encarnación Peña Ybarra (1918–1991).

WRITING THE GOODLIFE

INTRODUCTION

Defining Mexican American Goodlife Writing

IN THE SUMMER OF 2013, on the heels of President Obama's speech on climate change, television host, comedian, and political commentator Bill Maher wondered aloud about the environmentalism of "Hispanics" on his long-running HBO program *Real Time*. Maher interviewed researcher Anthony Leiserowitz, who investigates the climate change attitudes of Americans and directs the Yale Project on Climate Change Communication.[1] The two discuss an intriguing result:

> BILL MAHER: Some of your findings were very surprising. For example, you found that the ethnic group that was most concerned with the environment? Hispanics. Why is that?
>
> ANTHONY LEISEROWITZ: I wish I knew the full answer to that.
>
> BM: Well, you studied it! [*audience laughter*] That's why you're here, Doc! [*more audience laughter*]
>
> AL: Not enough! There's always a need for another study, Bill! [*smiling*]
>
> BM: You're throwing this back at me? [*laughing, hands in the air*]
>
> AL: [*getting serious again*] Well, partly we know that the people who are most concerned about climate change have what we call strong egalitarian values, whereas people who are the most hostile to the issue of climate change have strong individualistic values. In other words, they're deeply suspicious of government, government needs to get out

of our lives, out of our way, more individual freedoms and autonomy and so on.

With this exchange, Maher and Leiserowitz attest to but cannot explain what to them seems "very surprising": that Hispanics more than any other single ethnic group in the United States favor changes in current energy policies in order to address climate change.[2] Very little scholarly research or anecdotes exist to make sense of Hispanic or Latina/o and, even less so, specifically Mexican American concern for environmental issues.[3] Rather than reinforce the common belief "that global climate change is primarily a concern of only upper and middle class whites," the Yale report states that its results "strongly suggest that these assumptions are often wrong. In fact, minorities often support action to respond to this global threat at levels equal to or greater than whites" (Leiserowitz and Akerlof 18). In the two years since Maher and Leiserowitz's conversation, six more polls have revealed results similar to the Yale report.[4] Media responses to these poll results mirror Maher's: surprise and questions. Leiserowitz offers Maher an explanation by suggesting that environmentalists embrace egalitarian values while the less environmentally inclined champion individualistic values, making the vague suggestion that Hispanics might be motivated by egalitarianism. But Leiserowitz himself admits that Latina/o approaches to environmental issues cannot be fully explained by pitting egalitarianism against individualism. Meanwhile, an article in the *New York Times* suggests that Latina/o concern for climate change emerges from our disproportionate exposure to toxic contaminants (Davenport). Ultimately, these explanations still leave a lot of unanswered questions.

Writing the Goodlife dispels the mystery about Latina/o environmental views by turning to the largest part of the Latina/o population in the United States—Mexican Americans and Chicanas/os[5]—and unearthing a genealogy of values and practices maintained in Mexican American and Chicana/o culture. Media responses to poll results tend to emphasize Latina/o concern for environmental justice, but I want to expand the discussion. The Latina/o relation with the environment does not only consist of a reaction to environmental injustice. Latinas/os have sustained a reciprocal relation with the natural environment over many years and by means of unique values and practices, even in the face of environmental injustice. I look to the literature published between 1848 and 2010 to write a history of what I call "goodlife writing." Goodlife writing embraces the values of simplicity, sustenance, dignity, and respect. These are the four values I found to be consistent over this broad expanse of literature, and together they function to preserve mutually healthy relations among individuals and communities. The values in goodlife writing implicitly integrate the natural environment as part of the community, and thus cultivate a life-sustaining ecology for humans.

This book unfolds the story of goodlife writing by starting, in chapter 1, with the crisis of identity and sovereignty that Mexican Americans experienced immediately following the U.S.-Mexican War. The ensuing chapters observe how the Mexican American culture shaped its relations as a community and with the natural environment up to the beginning of the twenty-first century. I begin with the imperialist takeover of half of Mexico's territory because it so fundamentally marks the Mexican American relationship with land, in the minds of Anglo-Americans and Mexican Americans alike. Chapter 1, which discusses the writings of María Amparo Ruiz de Burton and Jovita González, sets the stage by showing how Anglo America rejected the knowledge Mexican Americans gained through generations of experience with the land in the new southwest United States. This resulted in a literal alienation of Mexican Americans from land and the natural environment as a whole. One might even call hypocritical the mainstream bewilderment in response to the Latina/o concern for environmental issues, for it was the power brokers of Anglo America who alienated Mexican Americans from the natural environment, especially by means of land grabs. Mexican American knowledge about climate and healthy land practices in the Southwest was largely lost in the latter half of the nineteenth century, discarded for the sake of dispossessing this community of new Americans. This tragedy lays the foundation for Mexican American goodlife writing. The following four chapters detail the values, practices, and culture that Mexican Americans cultivated in their relations with one another and with the natural environment, relations which I label "the goodlife."

The first aspect of the goodlife that I explore concerns the way colonization marks Mexican Americans' relationship to the land by means of common experiences of exploitation and objectification. Chapter 2 looks to the writings of six different authors who connect the colonization of Mexican American bodies and minds to the colonization of land. By the first half of the twentieth century, writers lament that Mexican Americans' access to land and power is proportionate to the degree to which they identify with Anglo-American culture and values. These Mexican American writers show how the experience of dispossession helps them understand the destructiveness of racial hierarchy and, important for alliance with the natural environment, the damage caused by objectifying land. This experience translates to the goodlife by instilling in Mexican Americans an ethic of sustenance in terms of natural resources, and an ethic of simplicity in overall lifestyle. After such experiences of humiliation and loss, Mexican Americans also cultivate dignity and respect with one another. These goodlife values and practices do not emerge fully formed or in a parallel fashion during this literary historical period, but chapter 2 notes their roots in these writings.

What happens in the latter half of the twentieth century is nothing short of extraordinary. Rather than conforming to the existing hierarchies and exploitations, Mexican Americans choose paths departing from mainstream values. They—especially the Chicana feminist writers—develop the idea to transcend possession, which exhibits significant intellectual and ethical independence from existing values. Rather than allowing dispossession of territories to comprise the center of their concerns, these writers reject the idea of possession itself as the bringer of misery, the practice that creates humiliation and subservience. Transcending the notion of possession sharpens the core goodlife values and characterizes one of Mexican Americans' most significant contributions to both human and natural relations today. Simplicity, sustenance, dignity, and respect all benefit from the general principle of transcending possession in regard to human-to-nature as well as human-to-human relations. In chapter 3 I delineate the concept of transcending possession as manifested in works by Chicana feminist writers and activists such as Enriqueta Vasquez and Elizabeth "Betita" Martinez, as well as writing about naturopathic healing by Rudolfo Anaya. Chapter 4 continues this thread and focuses specifically on writing about farmworkers by Tomás Rivera, Tish Hinojosa, and Helena María Viramontes. These writings range from engaging civil rights–era activism to representing the quiet power of alternative medicine and the dignity of laboring on the land.

By the beginning of the twenty-first century, Chicana feminist writers best articulate goodlife writing, including transcending possession. San Francisco Bay Area playwright, essayist, and poet Cherríe Moraga most explicitly engages ecological issues in her work, in particular a trilogy of three works I examine in chapter 5. In those three works—*The Last Generation*, *Heroes and Saints*, and *The Hungry Woman*—written and produced at the turn of the century, Moraga offers the value of care to replace destructive possession. What makes her work of stark interest is that she manifests care in a most extreme manner: she takes care for the human-nature relation to its logical conclusion by showcasing the ultimate surrender to simplicity, sustenance, dignity, and respect in the form of human sacrifice and death. Such an image musters visions of Aztec rituals atop dramatic pyramids, but Moraga's work is decidedly contemporary and relevant to twenty-first-century racial, sexual, and ecological politics. She draws attention to the death and destruction committed for the sake of possession and hierarchy. Corporations, oil explorers, and chemical manufacturers, to name just a few, sacrifice human life on a daily basis and on a grand scale, albeit shielded by laws that govern the trade's and profits' bottom line. Moraga shows how death can be a healthy part of an ecological process as well as an effective political strategy that draws attention to the assault on human life committed by the way we live today and the exploitative values that are taken for granted under capital, modernity, and coloniality. Giving up one's life, or the life of a loved one, when not for the sake of monetary gain, represents the ultimate act of transcending possession.

Despite this long history of goodlife ethics and practices, there is a reason why Mexican American and Chicana/o environmentalism remains a mystery: we never became environmentalists in the first place. While the American project of environmentalism denotes an explicit quest to find alternatives to exploitative approaches to nature, goodlife writing shows how the Mexican American and Chicana/o culture enacts values and practices that include nature all along.[6] Goodlife values can be found in Mexican American writings published as early as the late nineteenth century, at the same time that these writings maintain a connection to pre-Columbian practices and epistemologies.

I found the name for what I call this body of work—"goodlife writing"—in a book by Fabiola Cabeza de Baca. She describes her childhood on the family ranch in her 1949 book *The Good Life: New Mexico Traditions and Food*: "Life as I grew up and as I knew it as a home economist was rich but simple. People drew their sustenance from the soil and from the spirit. Life was good but not always easy" (v).[7] The core values that Cabeza de Baca lists—not too much but just enough, wealth measured by degrees of simplicity and community rather than material accumulation, a sense of respect for the dignity of the spirit and of the land all in one breath—resonate with the lives and approaches that I find in abundance in Mexican American and Chicano/a writing from the late nineteenth century to today. The notion of "the good life" also has a lot in common with movements in Latin America that successfully challenge conventional discourses of development. For example, in 2008 both Ecuador and Bolivia enfranchised the natural environment in their respective constitutions.

As described by anthropologist Arturo Escobar, "The Constitutions introduced a novel notion of development centered on the concept of *sumak kawsay* (in Quechua), *suma qamaña* (in Aymara) or *buen vivir* (in Spanish), or 'living well'" (Escobar 138). Escobar adds that this new conceptualization "subordinates economic objectives to ecological criteria, human dignity, and social justice" at the same time that "it represents an unprecedented 'biocentric turn,' away from the anthropocentrism of modernity" (138). The Mexican American and Chicana/o goodlife writing I document in this book embraces similar values and turns away from the modernity imposed during and after colonization. Viewing Mexican American and Chicana/o writing with the goodlife lens reshapes American literary history and environmental studies in a way that gives precedence to community- and dignity-oriented values that survive and thrive despite assaults that started during the colonization process.

For example, Cabeza de Baca's *The Good Life* has received little attention within American or even Mexican American literary studies. However, another book published during the same year is considered a classic of American environmentalism: Aldo Leopold's *A Sand County Almanac*. The concurrent publication of these two books offers an opportunity to see how Mexican American views concerning the environment have been in existence for as long as, if not longer than, mainstream U.S. environmentalism. Wallace Stegner, another one of the twentieth century's most renowned American environmental writers, salutes *A Sand County Almanac*: "When this forming civilization assembles its Bible, its record of the physical and spiritual pilgrimage of the American people, the account of its stewardship in the Land of Canaan, *A Sand County Almanac* will belong in it, one of the prophetic books, the utterance of an American Isaiah" (233). Stegner originates the ongoing label for Leopold's book as the "Bible of the contemporary conservation movement." In a deceptively simple narrative, Leopold tells of the passing of each month during one year. In describing each month, he also reflects on various challenges and questions facing the contemporary human relationship with our natural environment. *A Sand County Almanac* is still being read today all over the United States and in translation across the globe, while Cabeza de Baca's writings remain largely unnoticed.

One famous, the other virtually unknown, yet both writers promote comparable approaches to the relationship between humans and nature. Stegner and many others praise Leopold's critique of consumer culture and his creation of the "land ethic." In regard to consumerism, Leopold advised: "Nothing could be more salutary at this stage than a little healthy contempt for a plethora of material blessings" (iv). As an alternative to accumulation, he carefully argues for a different way of thinking about the relationship between humans and the natural environment. He calls his new idea "the land ethic": "Examine each question in terms of what is ethically and esthetically right, as well as what is economically expedient. A thing is right when it tends to preserve the integrity, stability, and beauty of the biotic community. It is wrong when it tends otherwise" (224–25). The core values found in Leopold's land ethic echo the core values found in goodlife writing. Not only did Leopold and Cabeza de Baca publish their books in the same year, but Leopold also spent some of his U.S. Forest Service years in Cabeza de Baca's homeland of New Mexico, where he also met and married his wife Estella Luna Bergere in 1912.

Estella hailed from a prominent Mexican American family in Santa Fe, New Mexico. Her mother's family, the Lunas, traced their heritage back to Spain and benefited from the generosity of the Crown in their acquisition of lands. They established a shepherding operation during the Spanish period and into the time of Mexican and then U.S. rule in New Mexico. The Lunas rose to prominence in the Southwest during the period following the 1846–1848 U.S.-Mexican War, and by the time of Aldo Leopold's arrival in their lives in 1911, the Lunas were an influential family at the center of political and social circles in Santa Fe. Ever since at least 2009, which was the centennial of Leopold's arrival in the Southwest, scholars have been considering the influence of Mexican and American Indian cultures on the development of his ideas.[8] Doubtless Leopold took in a great deal of the diverse approaches to the natural environment during his time in the Southwest. In *A Sand County Almanac*, Leopold's resonance with the goodlife continues: "A land ethic changes the role of *Homo sapiens* from conqueror of the land-community to plain member and citizen of it . . . it implies respect for his fellow-members, and also respect for the community as such" (204). Given some similar elements of provenance, the commonality between the land ethic and goodlife writing is not surprising. But consider the contrast in regard for the land ethic versus the almost complete lack of knowledge about a whole body of writing's (Mexican American and Chicana/o) contributions to environmental thought. Leopold's critique was timely, lyrical, and certainly deserving of its following. How much more impact could the land ethic and other approaches that challenge the domination of nature make if we bring them into the picture alongside one another? Goodlife writing advances values similar to those found in the land ethic, and it has been doing so since at least the late nineteenth century and in conjunction with a distinct culture's literary history. For these reasons, I take Cabeza de Baca's book as a chronological midpoint and the title for this literary history. It is time to stop pretending that ideas for healing the relationship between humans and the natural environment need to be new or originate in Anglo-American writings when they can be found in the heart of a thriving culture that has stared down assault and assimilation, and won.

GOODLIFE WRITING IN ECOCRITICAL CONTEXT

Goodlife writing enriches both ecocriticism and Chicana/o studies, pushing both fields into new territory. First of all, goodlife writing questions fundamental assumptions made by ecocriticism, thereby opening up this critical inquiry for more contributions from Chicana/o and Latina/o literature. More importantly, goodlife writing rectifies some destructive ecocritical practices, challenging the field to once and for all abandon its tacit approval of settler colonialism implicit in its first wave enthusiasm for the pastoral and the biocentric. Although it seems the second wave might better account for such limitations with its implementation of poststructuralist approaches, all but the environmental justice and postcolonial areas of ecocriticism succumb to the allure of devising global-scale theories that satisfy utopic inclinations but forsake the value of knowledge found in specific cultures aligned with caring for particular places, be those long-time or ephemeral homes. A brief review of the development of ecocriticism can help put goodlife writing into context, as well as highlight its contributions to this still relatively new field of literary inquiry.

We are still within the first thirty years of the formalization of ecocriticism. It is a rapidly changing field, and it is self-conscious of its development, consistently narrating its shifts. This is partially because ecocriticism remains a fundamentally generous field in which its critics readily engage with challenges and accept a change to their approaches if given appropriate reason. Another cause for ecocriticism's changes in such a short period is that as a new field, it is still locating its identity. In this regard, it is marked by its emergence during the postmodern era. It also prides itself on being at the place where the ivory tower and the activist march meets, so it remains attuned to shifts in popular culture and the ever-receding horizon of activist concerns. Finally, ecocriticism defines itself in relation to the science of ecology, and as knowledge shifts in that area of inquiry, ecocritics must also shift their approaches.

Generally, critics agree that there have been two initial stages of ecocriticism. We differ as to how the future will develop. Ecocriticism, simply put, is the study of the natural environment as it appears in literary and cultural productions. Scholars have been carrying out this kind of analysis for as long as literary and cultural studies have been written. However, with the establishment of the Association for the Study of Literature and Environment (ASLE), this field of study underwent a shift toward a group identity and toward pro fessionalization. ASLE was created in 1992 during a meeting of the Western Literature Association. The typical characterization of the first era of ecocriticism describes it as emphasizing nature writing and a literal, straightforward engagement with the natural environment. In a recent joint article, critics Lawrence Buell, Ursula Heise, and Karen Thornber describe it:

> First wave scholarship of the 1990s tended to equate environment with nature; to focus on literary renditions of the natural world in poetry, fiction, and nonfiction as means of evoking and promoting contact with it; to value nature preservation and human attachment to place at a local-communitarian or bioregional level; and to affirm an ecocentric or biocentric ethics, often intensified by some conception of an innate bond . . . conjoining the individual human being and the natural world. (419)

This description emphasizes the literal interpretations and the bioregional values of early ecocriticism, and I also note the absence of a significant engagement with the performing arts and other media in early ecocriticism. The early era of ecocriticism was also absent of diversity. Where ethnic diversity appeared, it was in discussion of American Indian literatures or cultures, and usually discussed as a phenomenon of a romanticized past rather than engaging with the American Indian cultures of today. The early era also showed an epiphanic and environmental reward for independence and isolation, as well as reinforcing the false separation of humans from nature. Critic Terry Gifford puts it more simply: during ecocriticism's "first phase American nature writing, wilderness literature and experiences of individual epiphany were respectfully celebrated with assumptions of simple realism" (Gifford 16). Yet another critic, Michael Cohen, derisively calls it the "praise song school" of ecocriticism (22).

As you can see from these few descriptions, there is a range of cynicism regarding this first stage of ecocriticism. Although I agree that ecocriticism has moved away from its earliest practices and concerns, I resist ridiculing it or making a harsh critique of it. I cannot bring myself to reject the early stages of a field of study that "begins from the conviction that the arts of imagination and the study thereof—by virtue of their grasp of the power of word, story, and image to reinforce, enliven, and direct environmental concern—can contribute significantly to the understanding of environmental problems: the multiple forms of ecodegradation that afflict planet Earth today" (Buell, Heise, and Thornber 418). If we keep in mind these roots of ecocriticism, which emphasize the power of human imagination, then we need not introduce a divisive understanding of its beginnings.

Still, we do need to remain attentive to the critical approaches that do not include the real challenges we face. I would like to introduce the idea that the development of ecocriticism can be understood through the idea of the decolonial. I use the term decolonial as it is discussed within Latin American cultural studies. First, the decolonial is not the same as the postcolonial. Neither are these two in competition with or contradiction to one another. As decolonial theorist Walter Mignolo states in his recent book *The Darker Side of Western Modernity: Global Futures, Decolonial Options*, "Both [postcolonial and decolonial] projects strive to unveil colonial strategies promoting the reproduction of subjects whose aims and goals are to control and possess" (xxvi). However, critics such as Mignolo, Anibal Quijano, and María Lugones still saw fit to develop another theory for the response to and survival beyond the colonial project. The decolonial emphasizes the cultures and peoples that have not only survived the colonial experience, but have also persevered with their epistemologies intact, as much as possible, despite the colonial project. This too may sound similar to the postcolonial. The primary distinction lies in the foundations of each—the postcolonial is genealogically based on Western theory, taking as a point of departure a modern subject for whom "knowledge can be objective" (Mignolo, *Darker* xxiv), while the decolonial prioritizes a non-Western theoretical basis and puts the body, a body politics of knowing, at its center. The poststructural/postmodern theories, from which postcolonialism departed and with which it still holds a relationship, center on abstract ideals and to a certain extent a universal humanism, while Mignolo asserts that "it is from the body, not the mind, that questions arise and answers are explored[; w]hat calls for thinking is the body not the mind" (xxiv). In short, decoloniality is a practice of "disengaging and delinking from Western epistemology" (xxv). Accordingly, we can trace the idea back to the Bandung Conference of 1955, where twenty-nine nations from Africa and Asia—nations that had recently taken a stand against colonization—met in Indonesia to reject the options of both capitalism and communism, as well as to resist the imposition to choose sides during the Cold War. They came up with the idea of the decolonial to describe the process they were going through toward claiming their own identity and system of governance.

In the context of the ecocritical and the environmental in general, three fundamental concepts underlie the decolonial disengagement with Western epistemology: (1) rejection of the idea of commodification, (2) refusal to pursue epistemological dominance—the decolonial is merely an option—and (3) a "geo- and body-politics of knowledge (understood as the biographic configuration of gender, religion, class, ethnicity, and language)" (Mignolo, *Darker* 9). If we consider the development of ecocriticism in relation to the decolonial, the early phase of ecocriticism would be its modern phase, when it understood nature as other than human, when it took the idea of wilderness as an undisputed good, and when it took for granted the privilege of biocentrism. Then, the ensuing developments can be categorized as occurring in at least two different branches: the poststructural/postmodern and the decolonial.

The second era of ecocriticism not only takes seriously the human relationship with the environment but also concerns itself with the way that human-to-human relations are negotiated via the natural environment and how the exploitation of humans as well as nature is often rationalized in overlapping ways. They also integrate an idea of nature as a constructed concept, rather than accepting it as fact.

In short, the second era embraces the social justice aspect of environmental studies. This means, of course, the inclusion of more racial and ethnic diversity. Diversity within ecocritical studies includes many areas such as race and ethnicity, sexuality, gender, and disability. Indeed, many scholars work to document the contributions of these various groups to environmental thought.[9] As my work is primarily concerned with racial and ethnic diversity—Mexican American literature, to be exact—I will focus my comments on that aspect, although I remain attuned to gender, sexuality, and ability as well.

In one of the earliest collections of ecocriticism (*The Ecocriticism Reader*, 1996), Cheryll Glotfelty called for ethnic and racial diversity in particular. However, this did not materialize in any substantive way until the publication of two books: *American Indian Literature, Environmental Justice, and Ecocriticism: The Middle Place* by Joni Adamson in 2001, and *The Environmental Justice Reader: Politics, Poetics, and Pedagogy*, edited by Joni Adamson, Mei Mei Evans, and Rachel Stein in 2002. These books both emphasize environmental justice as an overlapping concern with social justice and environmental issues, the first concerning American Indian environmental writing, and the second a collection of essays on many aspects of diversity within ecocriticism. Both books made early interventions into ecocriticism's largely straight Anglo-American male-centered explorations of nature writing. For example, Adamson writes of her experiences learning from present-day American Indian student populations as well as the literatures they read together, avoiding the objectification of American Indian culture as a thing of the past. These books also clearly established that much work remains to be done to fill the gaps.

We are now in the middle of an explosion of work that responds directly to these calls to action. Books like *Race and Nature* by Paul Outka; *Black Nature: Four Centuries of African American Nature Poetry*, edited by Camille Dungy; and *Shades of Green: Visions of Nature in the Literature of American Slavery* by Ian Finseth have all been published within the past five years. There has also been a special issue of the journal *MELUS* that addresses diversity in ecocriticism, and countless articles and conference presentations that perform this work. This means that interventions are being made, and ecocriticism is slowly considering more and more diverse authors.

We are also seeing ecocriticism explicitly engaging more and more with critical theory, whereas its earlier stage emphasized primary texts, and we are seeing a growth in particular with postcolonial ecocriticism. Ursula Heise, Tim Morton, Catriona Mortimer-Sandilands, Rob Nixon, Elizabeth DeLoughrey, and George Handley are just a few of the critics who are leading these conversations. One of the more promising developments has been the argument to move away from humanism—the posthuman—from critics such as Cary Wolfe. I would like to make a distinction between the way I discuss human bodies existing as a part of nature as opposed to the humanist, the Enlightenment/modern concept of human intellect and transcendence. Posthumanist Cary Wolfe proposes we think of "humanism as a historically specific phenomenon" ("Introduction: What Is Posthumanism?"). In other words, our present way of thinking of humans has not always existed, and we are beginning to see the end of this concept's epistemological dominance. In this approach, posthumanists put us in the position of rejecting the aspects of humanism that have proven destructive—for example, the mind/body and the human/nature dichotomies. I agree with this approach up to that point, but I tend to see a radical questioning of humanism itself as a better option, while Wolfe still holds to some aspects of humanism. In his book *What Is Posthumanism?* he says, "The point is not to reject humanism *tout court*—indeed, there are many values and aspirations to admire in humanism—but rather to show how those aspirations are undercut by the philosophical and ethical frameworks used to conceptualize them" ("Introduction: What Is Posthumanism?").

Again, I agree with the idea that not every aspect of humanism is without value, but I still want to approach our current ecological crisis with a more radical proposition of engaging the ideas and processes that have not succumbed to modern values. In a word, this approach can be called decolonial. In the remainder of this introduction, I propose the decolonial as an important new means of cultivating environmental approaches that allows us to avoid our destructive humanism. In short, the decolonial involves a rejection of Western epistemology in which we break down the dichotomy between humans and nature and make space for indigenous practices and narratives that have survived colonization and that preserve and adapt traditional environmental knowledge.

With the decolonial approach, we can also finally make sense of the perceived lack of interest in environmental issues on behalf of the culturally marginalized. For a long time, environmental activists, as well as academic studies such as ecocriticism, environmental history, and environmental philosophy, have lamented their lack of diversity. They asked themselves, and still do, how can we better appeal to the lower classes and the ethnically diverse peoples to care for the environment the same ways that we middle- and upper-class and ethnically mainstream people do? In the United States, it is the middle- and upper-class Anglo-American population that has always seen itself as taking the lead on environmental reforms and innovations, such as the establishment of national parks and the management of natural resources such as forests. Partly understanding this disconnect between environment and the marginalized, some scholars (myself included) have tried to make connections to the environmental movement with the idea of environmental justice. Environmental justice, with its central concern for human access to nontoxic living and working spaces, as well as to clean water and healthy food, has made some progress toward connecting the marginalized with environmental concerns. This is the "social justice transformation" of the second wave of ecocriticism. But this concern for humans is still not enough.

Colonized culturally and economically, marginalized peoples like Mexican Americans do not find conventional environmentalism appealing. Scholar Randy Ontiveros sharply observes, "The reality is that while environmentalism claims to speak for all of humanity, in practice it has been rooted in the politics of race and power" (Ontiveros 90). *Writing the Goodlife* shows how Mexican American literary history best contributes to environmental literary studies by undermining the category of environmentalism itself. In assembling this literary history from 1848 to 2010, I attend to the ways that the culturally marginalized never wholly bought into the modern worldview that centers on nature/culture dichotomies. As such, they do not embrace environmentalism as moderns and even postmoderns understand it, given that the postmodern is positioned dialectically in reaction to the modern.[10] The economically impoverished, the colonized, and the culturally marginalized by necessity ensure their own survival in terms of jobs and education, food and water, but they do not define or understand their actions as "environmental." As Walter Mignolo argues in *The Darker Side of Western Modernity*, within decolonized cultures, "there was not, and there is not today, a distinction between 'nature' and 'culture'" (11). They have always been among the most exposed to environmental exploitation—manual labor, toxic contamination, sovereignty loss—but they see this as a product of modern epistemologies that justified colonization of peoples and places. As a result, they did not have to develop an environmental critique to solve the nature/culture divide: their approach is at once nonmodern and decolonial, but it still offers environmentalism an alternative set of traditions and insights with which to approach today's challenges.

Although not self-identifying as environmentalists, Mexican Americans remain the people for whom environmental degradation is most relevant because they are among the most vulnerable to the consequences of environmental destruction.[11] If we do not make visible Chicanas' and Chicanos' unique ways of negotiating environmental issues, our silences encourage misrepresentations and ultimately foster further exploitation. The advantage of learning from goodlife writing is at least twofold: because Mexican Americans trace their heritage back to the original colonial encounter, environmentalism can learn from an epistemology that has withstood colonial assault for over five hundred years, and the decoloniality of Chicana/o studies is strengthened by showing how it has also resisted environmental dispossession and degradation all along. The perplexity evident in the exchange between Maher and Leiserowitz on *Real Time* only begins to suggest how important it is to build a bridge between Mexican American literary history and environmental studies and to expand the reach of ecocriticism.

Goodlife writing does not explicitly engage environmental issues or follow the typical sequence of events highlighted in other Chicana/o literary histories. Rather, goodlife writing offers insight into this culture's unique approaches that include fostering a healthy relationship among humans and with the natural environment. For example, goodlife writing rejects the concepts of possession and control. This is in contrast with conventional environmental and Chicana/o values. The typical genealogies for both these areas include concern for possession and control of territories—environmentalists in the name of preservation or conservation, and Chicanas/os for economic independence and continuity of settler-colonial heritage. However, the genealogy I trace for goodlife writing suggests that some Mexican American and Chicana/o writers and activists, particularly women, have detected a destructive vein in the desire for control and possession of lands.

Given that so many of these works have been penned by women, goodlife writing brings women's voices and concerns to the forefront. Mexican American women's and Chicanas' writings focus on their lived experiences that tell them how patriarchy, domination, possession, and control oppress women at the same time that they oppress the natural environment. The Chicana emphasis on experience also demystifies individualistic rights while it dignifies the many interests and concerns that make up everyday life, which they consider a legitimate and powerful site of knowledge production. This departure from individualistic rights also shifts the focus from justice based on equality and toward justice based on dignity and respect. The release of control and domination also creates a home for the full spectrum of sexualities and genders—something with which both Chicana/o and environmental studies have struggled.

EXAMPLES OF GOODLIFE WRITING

What are some specific examples of goodlife writing? In his *Black Mesa Poems*, Jimmy Santiago Baca writes an ode to the spring snowmelt that fills the *acequias* (irrigation ditches). Goodlife writing celebrates traditional and environmentally sustainable agricultural practices in northern New Mexico.[12] In

capitalist modes of nature objectification. In addition, this insight shows how an environmental analysis adds a new dimension to a dominant understanding of Mexican American literary production and history. Accounts of the Chicana/o Civil Rights Movement focus on the efforts to regain control of lost lands, and to their detriment they miss other voices that were just as creative and committed to justice. We need to also attend to the ways that Chicana/o writers teach us to look beyond mere possession of territories and goods to promote anticonsumerist values (Rosales 154–70). In the late twentieth and early twenty-first centuries, Chicana feminist writers continue the legacy of the Civil Rights Movement's rejection of possession and seek to preserve a relationship with nature built on the values of respect and autonomy.

And, finally, if Mexican American writing brings attention to environmental injustice, is its point of view concerned primarily with critique rather than offering alternatives or solutions to environmental crisis and climate change? Do Chicanas/os appear in an environmental context only as victims of exploitation, or can they also be a source of knowledge and alternative approaches? A great deal of this body of work brings attention to environmental injustice, but so much of Chicana/o literature also testifies to alternative, decolonial environmentalisms evident within Mexican American culture—a treasure trove of knowledge virtually unknown to and unrecognized by environmental studies.

Not as interested in the rhetoric of rights and equality emphasized by environmental justice, these decolonial environmentalisms embrace dignity and respect as their core values. Blending the practical and the theoretical, decolonial environmentalisms range from the popular practices of *curanderismo* (folk medicine) and hybrid spirituality to backyard *milpas* (corn fields) and recycling—by necessity and convention rather than in the name of environmentalism. In order to answer the question, "What is Mexican American goodlife writing?" a long history and a whole range of genres, styles, and eras must be considered. In order to understand and value Mexican American contributions to environmental thought, and the key to this long history and literary diversity, one needs to see the interplay of land dispossession with this culture's transcendence of the ideas of control over nature and determination to preserve dignity and respect. I call this process "transcending possession," as it rejects the notion of control and it cultivates respect among humans and the natural environment. It also rejects the objectifying tendencies of rights- and private property–oriented modernity.

Indeed, transcending possession best describes the contributions made by the widely varied works of Chicana/o goodlife writings. These writings add much-needed diversity to conventional environmental approaches, while at the same time upending one of the most stereotypical ideas about what concerns Chicana/o thought. Chicana/o studies and politics assume a cultural predisposition toward taking back the lands lost after the U.S.-Mexican War, either literally or figuratively. The concept of Aztlán best exemplifies this desire. Aztlán is the imaginary homeland that young Chicana/o activists wrote about and rallied around during the Civil Rights Movement, and it still inspires both creativity and consolation today. Geographically, Aztlán consists of an area in the present-day U.S. Southwest, the original home of the Aztecs before they migrated to central Mexico in the twelfth century and established the empire that they ruled from the city engineered from a swamp, Tenochtitlán. Figuratively, Aztlán is more difficult to define. When, during the Civil Rights Movement, Mexican American youth felt themselves without a territory—not at home in either the United States or Mexico—they turned to Aztlán as an imaginary homeland, an ideal place of belonging, that has been a keystone of Chicana/o cultural production ever since. Aztlán exists wherever a Chicana/o individual or community dwells and embraces the ideals of dignity, sovereignty, freedom from prejudice, and opportunities for self-determination. Aztlán has helped many Chicanas/os feel a sense of belonging, if not to a particular territory or nation, then at least to a cultural imaginary.[15] The fact that it references a historical and physical territory can imply that Chicanas/os actually aspire to repossess lost lands. However, I argue that Chicanas/os' original trauma of alienation from their homeland eventually leads many to envision transcending actual or figurative possession of land and reclaiming their own dignity as individuals and as a community. Chicana voices have led the way toward this transformation, largely due to how they sought to create a space for themselves within the Chicana/o identity that took shape during the Civil Rights Movement.

One Chicana who leads the way is Cherríe Moraga. As Mary Pat Brady shows, Moraga transforms the significance of Aztlán for Chicanas/os who identify as biracial and/or queer and/or female: "Moraga travels from an (anti) origin in cultural nationalism that rendered her alien because of her impure status to a position within an imagined homeland in which the intersections of subjectivity might not only be taken into account but be fully explored" (150). Moraga takes the original idea of Aztlán, which was so closely tied to repossessing the lands in the name of nationalism, and turns it into an imaginary homeland that does not require material tangibility or possession. She finds she must make this transformation because, while she identifies with the idea of Aztlán, she cannot inhabit it unchanged as a biracial lesbian who refuses to endure male dominance. The new Aztlán that Chicanas envision, empowering for the entire spectrum of sexualities and genders, by definition resists possessiveness and domination while it works toward dignity and respect.

EARLY CRITICISM OF CHICANA/O WRITING AND THE ENVIRONMENT

The earliest critics to place environmentalism alongside Chicana/o writing, Kamala Platt and María Herrera-Sobek, both center their analyses on Chicana feminist writings from the 1990s that emphatically cultivate dignity and respect while also transcending possession. These first landmark efforts toward documenting Chicana/o environmental contributions center on Chicana feminist works that best distill the Chicana/o dialectic between dispossession and the rejection of possession itself, basing their denunciation of control on a critique of patriarchy alongside racism, homophobia, and capitalism. Kamala Platt, in the earliest published ecocritical analysis of Chicana/o writings in 1996, argues that, in Ana Castillo's *So Far from God*, the ecofeminist alliances forged in resistance to capitalist attempts to control agriculture in northern New Mexico "strive to attain self-determination and sustainability and to retain a symbiotic relationship between the needs of the people and the needs of nature" (Platt 85). This echoes a real-life cooperative documented by Laura Pulido and shows how the ecofeminist Chicana ethic denounces the exploitation of both women and the natural environment as well as emphasizing communal stewardship rather than individualistic possession (Pulido 1996). María Herrera-Sobek, in another early contribution to Chicana ecocriticism in 1998, argues that Chicana writers consistently work toward "the constructions of new ethical systems for the Chicano/a population," and she shows how an antipatriarchal approach to human-nature relations is inherent in these newly designed ethical systems (Herrera-Sobek, "The Nature of Chicana Literature" 91). She further describes her approach: "My analysis posits that the Chicanas' critical discourse on ecological issues emanate from gender, ethnic, and social perspectives" (89) and argues that "Chicana writers have been at the forefront of feminist ecological concerns since the early 1970s" (90). Thus, Herrera-Sobek shows how Chicana feminist ecological views are informed by lived experience rather than an appeal to an essentialized racial and gender identity.

More recently, critic Jorge Marcone describes Chicana feminists Gloria Anzaldúa and Cherríe Moraga as "the most influential Chicana writers at the turn of the [twenty-first] century," whose work "pays attention to the cultural construction of nature (including within environmental movements) and the marginalization or exploitation of women, gay men, lesbians, ethnic minorities, etc., that such constructs support" (Marcone 195). This analysis, resonant with those of Platt and Herrera-Sobek, points to the way Chicanas' experiences inform their critiques of discrimination, and they refuse to exclude the natural environment from the list of entities that endure invidious cultural constructions. Marcone adds that "Anzaldúa and Moraga intermesh the social and the cultural with the more-than-human world by weaving race, class, gender and sexuality, and migration with ecological issues" (195). By further unveiling Chicanas as the leaders of Mexican American environmental writing, my book participates in the current recovery of Chicanas' formative impact on the Chicana/o Movement and on how the period itself led to some of the most innovative work. Maylei Blackwell details this recovery in her book *¡Chicana Power!: Contested Histories of Gender and Feminism in the Chicano Movement* and in particular shows how "many of the theoretical innovations attributed to women of color feminisms of the 1980s, such as the concept of intersectionality or interventions regarding multiple subjectivity ascribed to the postmodern turn in feminist theory, in fact have their roots in the political views of women of color activists in social movements of the 1960s and 1970s" (Blackwell 2). These Chicana feminist concepts of intersectionality and multiple subjectivity from the 1960s still apply today and reinvigorate environmental approaches, especially in the way intersectional and multiple subjectivity contributes to transcending possession by demystifying individualistic rights and dignifying the many interests and concerns that make up all of our lived experiences.

In fact, Mexican American writers have never strayed far from confronting the environmental challenges that expand the confines of conventional environmentalism. Each era of Mexican Americans' historical struggle for recognition and autonomy since the mid-nineteenth century in fact concerns a range of environmental issues. However, Mexican American writings have rarely been considered environmental by ecocritical literary studies because their style and content do not match conventional nature writing, or because ecocritics by and large do not specialize in Chicana/o literature and culture.[16] For its own part, Mexican American literary study has rarely considered its works environmental because the human relation to nature is so fundamental to nearly every work that it becomes impossible to distinguish from other concerns. Additionally, Mexican American literary study remains consistent with its community's alienation from conventional environmentalism. For each work that I label "environmental," one can find an alternative and appropriate category of inquiry: social justice, identity, gender, sexuality, and more. This mutual disconnect between Mexican American literary studies and ecocritical studies reveals a fundamental challenge to bringing these two fields together: to what extent can we or should we consider Mexican American writing environmental?

Another way of asking the same question: to what extent can colonized peoples and places and the resulting hybridized Western/indigenous cultures make reference to "the environment" in the same way that the colonizing, dominant, and putatively unitary cultures do? How useful is the idea of "environment" in a Mexican American context? On one hand the relevance of environmentalism to Chicana/o literature is clear. Take, for example, Cherríe Moraga's environmental justice writings, or the way Jovita González's folklore redefines "nature writing," or the environmental history that María Amparo Ruiz de Burton's novels teach us. On the other hand, for these works to make sense environmentally means a transformation of each field. Ecocriticism must take a hard look at its elitist history and factor the legacy of colonization and racism into its selection of texts for analysis and critical stances toward those texts. Kimberly N. Ruffin makes a similar observation as she documents African American environmental voices in her recent study, *Black on Earth*: "Critical guideposts based solely on the experiences of European Americans neglect the experiences of negatively racialized Americans who also have struggled for connection with nonhuman nature and are therefore insufficient in illuminating African American ecological perspectives" (13). As Ruffin says, we need an alternative genealogy in order to appreciate the contributions of "negatively racialized Americans." At the same time, ethnic American literary studies can no longer relinquish environmental concerns to the limited sphere of the way it has been studied in the past. Ethnic American literary studies must begin to take full advantage of the way an environmental lens calls attention to new ways of historicizing and contextualizing both the losses that our cultures suffered as well as the knowledge we have preserved in the realm of human and ecological relations. In their introduction to the 2005 collection *Caribbean Literature and the Environment*, Elizabeth DeLoughrey, Renée Gosson, and George Handley observe, "Unlike the white settler production of nature writing, Caribbean writers refuse to depict the natural world in terms that erase the relationship between landscape and power" (4). In order to understand and appreciate the views of writers that emerge from colonial histories and colonized places, power dynamics must factor into the equation, and these editors make the point that for nature writing, landscape in the context of environmentalism has not historically been understood in terms of the full range of human-to-human power hierarchies.

If we really want to make evident the dynamics of human-to-human hierarchical power, such as race, class, gender, and sexuality, which impact our relationship with the natural environment, should we use the term "environment" in our studies of the views of colonized peoples and places? Or, do we understand nature in such a different way that we merit a different term? Once we consider Chicana and Chicano literary works and the insights they offer as strategically decolonial, to what extent does the term "environment" remain relevant? What does this tension between decoloniality and environmentalism produce? Something I call "goodlife writing."

DECOLONIALITY AND GOODLIFE WRITING

In fact, one benefits most by seeing the ecological lessons of Chicana/o cultural productions through a decolonial lens, which *Writing the Goodlife* implements. The decolonial, according to Walter Mignolo, is the response and alternative to the European modernity that went hand-in-hand with colonization. The decolonial claims Latin America's locus of knowledge in its own geography, place-time, and peoples (indigenous and mestizo/hybrid) as an alternative to "the idea that European modernity was the point of arrival of human history and the model for the entire planet" (Mignolo, *Darker* xiv). Mignolo further explains that the "darker side of modernity materialized in this belief. I explain it as 'the logic of coloniality'" (xiv). I describe goodlife writing as decolonial because it embodies two core values of decoloniality: (1) a consistent rejection of the modern ideology of universal humanism and linear progress, and (2) a deviation from chronological and single-dimensional approaches to time and place. Certainly environmentalism can learn a lot from the decolonial epistemologies that have endured in the pages of Chicana/o goodlife writing and survived the processes of colonization and its ensuing transformations. But to what extent is it appropriate to say that Chicana/o writings offer us environmental insights? We should instead argue that Chicana/o writings offer ways of thinking that do not require the legacy of modernity that accompanies coloniality and brought about the destruction that called for environmentalism in the first place.

Despite the rapid growth of both the field of ecocriticism and the field of Chicana/o literary studies, a dissonance has persisted between Chicana/o literature and ecocriticism that has kept us from developing a robust collection of Chicana/o ecocritical studies. This dissonance has to do with a problem of contrasting genealogies for Chicana/o studies and ecocriticism, both of which emerged from social justice movements. Put into the context of coloniality, the challenge of contrasting genealogies becomes clearer. This literary history of Mexican American goodlife writing brings together two disparate fields: Chicana/o studies and environmental studies. It would seem to be an easy connection to make; both fields work on behalf of disenfranchised, disempowered entities: ethnic minorities and the natural environment. Still, this is a seemingly unlikely book.

I initially ask the question, "What is Mexican American goodlife writing?" for the simple reason that I have had to answer that question many times, with the inquirer often implying that I will have a hard time convincing him or her that such a thing exists, and that a literary history of Mexican American writing about the environment sounds particularly fishy coming from someone as urban as I am and writing about a literature consistently discussed under the rubric of social justice, civil rights, and ethnic identity. The literature historically associated with environmentalism concerns a genre called "nature writing." Works that fall under the category of nature writing often tell of an individual's lonesome and reflective retreat to live in a cabin, hike in the mountains, or raft down a river. Think of such classics as Henry David Thoreau's *Walden; or, Life in the Woods*, John Muir's *My First Summer in the Sierra*, or Henry Beston's *The Outermost House*. These works offer many valuable lessons for healing the human relationship with the natural environment. However, having limited our understanding of environmental literature to this category of works makes it difficult to see how Mexican American literature fits in. Mexican American writings usually concern the everyday, where we live, where we work, and often include a crowded array of characters rather than a solitary soul on an isolated retreat. The urgency of our current climate crisis requires that we all consider as many ways of thinking about the environment as possible. Taking account of Mexican American goodlife writing does not just offer an expanded reading list and new insights, it also requires ecocritics to proactively regrind the lens through which we examine literature in order to discern its contributions to environmental debates. It also asks Chicana/o studies to reexamine its genealogy, especially in regard to the idea of transcending possession and the centrality of Chicana thought in developing this concept.

Why has it been so difficult to make this connection? What gets in the way? My response to this question speaks to a specifically U.S. context. In the United States, the popular representation of the Civil Rights Movement's arguments against racial denigration are understood to appeal to the values of universal humanism that emphasize our commonalities across difference. This popular understanding of racial equality has motivated generations of activists and prompted improvements in racial relations in the United States. Even so, we can benefit from putting humanistic values into a global historical context. These supposed universal values emerge from the modernity that co-constituted colonization. In other words, working against racial denigration succeeds in exposing the racism inherent in colonization but falls short of a robust critique of modernity. This leaves the Civil Rights Movement operating from an internally conflictive frame of reference; humanism never meant to recognize peoples of color as human. On the other hand, U.S. environmentalism has deconstructed the exploitation of nature by looking to and rejecting modernity (the human/nature, culture/nature, mind/body false dichotomies), but U.S. environmentalism has largely ignored the coloniality that accompanied modernity.[17] Without recognizing coloniality, environmentalism misses a major force that works to objectify nature. At the same time, popular U.S. environmentalism tends to emphasize universal humanism in its efforts to convince Americans to see our common plight of environmental crisis, too easily ignoring the fact that toxic contamination and other environmental assaults disproportionately impact populations of color and low income. And in response civil rights activists demand environmental justice. Once again, another major social movement of the twentieth century, environmentalism, operates from an internally conflictive frame of reference. These popular notions of civil rights and environmentalism undermine an awareness of the interdependence of coloniality and modernity in both the creation and continuation of racism as well as the concomitant degradation of the natural environment.

Although ethnic studies and environmental studies have developed more nuanced understandings of each of these dynamics, the initial disidentification still impacts the ways these two fields understand one another. This book is an effort to bring them into conversation. If each field recognizes the destructive roles of coloniality along with modernity today in terms of racial hierarchies alongside environmental exploitation, then this inquiry can dig down to a more fundamental level of understanding. Simply put, this means entering the intimidating territory of affirming differences rather than commonalities, but the time is right to move away from the utopian ideal of a universal humanism, especially within environmental thought that so easily lends itself to the global scale at the expense of local peoples and places. This is why I call the writings about nature by Mexican Americans "goodlife writing" rather than environmental writing. These goodlife writers affirm a decolonial disengagement from a conventional understanding of environmentalism.

Consider Gloria Anzaldúa's statement in *Borderlands*: "this land was Mexican once, / was Indian always, / and is. / And will be again" (113). In this statement she claims a Chicana/mestiza alliance with the land, speaking of its "identity" as she does her own identity in other portions of the book, emphasizing her shared fate with the land rather than an idea of her domination over it or even submission to it. In this quote we might consider "land" an equivalent concept to environment—both terms emphasize the "outsideness" of humans from our surroundings. Anzaldúa insists that not all peoples accept the human/nature dichotomy introduced to Western ideology during the Enlightenment. Taking just this statement into account, one can begin to see that no matter how much environmentalism struggles to reunite humans with our natural environment—to undo Western modern epistemology, to heal the false dichotomy between humans and nature—the term "environment" itself reifies this separation. In short, one of the most significant things that environmentalism can learn from the decolonial writings of Chicanas and Chicanos (among other peoples of color) is the fact that we never needed to become environmentalists in the first place, and we therefore have an array of strategies at our disposal for how to live well with Earth.

Lamentation over the paradigm of modernity is a familiar refrain within environmental studies. Still, some dispute this outright indictment of modernity.

Political theorist Andrew R. Murphy explains, "The antimodern narrative of environmental decline views instrumental rationality (the Hobbesian view of human reason as concerned only with means to ends) and mechanical philosophy (the Cartesian view of nature as passive matter in motion) as the enduring legacies of the scientific revolution" (86). Murphy argues against the idea that modernity is responsible for environmental decline by offering exceptions and alternatives to the above-described purely instrumentalist and rationalist legacies of the Enlightenment, and he does this using Western intellectual history. For example, he explains that "Ernst Haeckel called for an accurate understanding of 'the true place of man in nature,' deriding the Christian and anthropocentric traditions of his day" (88).

In the same vein, Timothy Morton, in *Ecology Without Nature*, notes that "[Theodor] Adorno based much of his work on the idea that modern society engages in a process of domination that establishes and exploits some thing 'over there' called nature" (8). Morton's much-cited book goes about a systematic deconstruction of the concept of nature in the name of advancing ecocritical thought. He attempts to keep ecocriticism honest, so to speak, and to advance it beyond its nature-writing focus. He aims to keep ecocriticism from reifying the very structures it works to transcend. Still, he does not see the full potential of decolonial goodlife values, which include valuing a relationship with the local:

> "Third World" environmentalisms are often passionate defenses of the local against globalization. Simply lauding location in the abstract or in the aesthetic, however—praising localist poetics, for example, just because it is localist, or proclaiming a "small is beautiful" aestheticized ethics—is in greater measure part of the problem than part of the solution. Our notions of place are retroactive fantasy constructs determined precisely by the corrosive effects of modernity. Place is not lost . . . We would be unable to cope with modernity unless we had a few pockets of place in which to store our hope. (11)

I agree with Morton's conclusion here—place is not lost—but I cannot embrace the reasoning that gets him to this conclusion. His reasoning includes the negation of any positive contributions by decolonial (or, as he puts it, "Third World") environmentalisms, calling their localism a "retroactive fantasy" (11) invented by modernity. He remains blind to the fact that some peoples never lost place at all, and it is for this reason that one can be appreciative rather than derisive of decolonial environmentalisms.

In the same vein, scholar Greg Garrard observes that environmentalism's much-lauded "Ecological Indian" is an invention of the West more than it is based on an actual past. In his book *Ecocriticism*, Garrard warns against looking to a colonized culture for its wisdom concerning how to dwell in place because "many of the indigenous societies whose knowledge and lococentric values bioregionalists admire are already thoroughly deracinated" (119). This argument verges on a second colonization that erases the knowledges that indigenous peoples have kept alive and adapted to the contemporary world. Garrard goes on to argue that any environmental knowledge found in indigenous cultures is more an invention of the West than authentic to the Americas: "Yet the idealization that would make Indians and other indigenous people models of ecological dwelling arguably derives primarily from [Euro-American culture], not the [Native American] culture" (123). He argues that the image of a Native American's proximity to nature owes more to eighteenth-century primitivist theories out of Europe than from anything based on historical reality (124). Again, I do not disagree that colonization led to a thorough attack on indigenous ways of being and knowing, but recognition of this history cannot then lead us to reject the environmental knowledge of indigenous peoples who survived the colonial assault. Both Morton and Garrard get this wrong because they allow a Western approach to chronology and epistemology to blind them. They refuse to see the still-living indigenous cultures that have so creatively coexisted with, but still defied takeover by, Western ways of knowing. Place is not lost because Morton recovers it or because Garrard credits Rousseau with inventing it, but because decolonial thinkers, dwellers, and activists have maintained and adapted their sense of place from the earliest days of colonization to the present day. They do this not by global means of a nostalgic response to "the corrosive effects of modernity" but through hard-won, day-to-day battles to keep up their knowledge practices. We need only look at some of their cultural productions to see their strategies and insights. Just as critic Theresa Delgadillo argues about hybrid spirituality in contemporary Chicana narratives, the goodlife writings that tell a story about human-nature relations apart from the colonial/modern version do so to "enact an imaginative partial recovery and recirculation of indigenous knowledge that function as countermeasures to hegemonic interpellation rather than as a nostalgic yearning for premodern simplicity" (Delgadillo, *Spiritual Mestizaje* 43).

I support Murphy's and Morton's, and in some ways Garrard's, investigation into alternatives to the instrumental and rationalist modernity within the Western tradition. Still, any alternative that emerges within Western modernity does not in any way challenge the decolonial analysis that supports Latin America's own geopolitics of knowledge. The decolonial is a parallel and fundamentally different narrative from the destructiveness of modernity and coloniality that sought to uproot the local ways of knowing beginning in the sixteenth century and that continue today. Ecocritic Patrick Murphy makes the point that current critiques of modernity from the perspective of postmodernity is a totalizing narrative that ignores those who never opted into modernity in the first place. He offers many ecofeminist writers as examples of this argument, including Chicanas Anzaldúa and Pat Mora, and he calls this approach "nonmodernity":

> What is ignored by such oppositions [of modernity with postmodernity] is the continuation of a nonmodernity—including various paramodern formations—that cannot be defined by the parameters of postmodernism . . . [M]any cultures in former colonies and current colonies refuse to accept *modernity* according to the economic, political, and cultural models used by the United States and the former colonial powers of Europe. They also resist the teleological conception of progress that models of modernity embody. As a result, they may prefer to be nonmodern on their own terms rather than modern on someone else's terms. (90)

The dispute is not with the whole of Enlightenment epistemology, but with the ways that any totalizing ideology threatens the persistence of local and disempowered ways of knowing. This book makes a humble attempt to document one strain of knowing: Mexican American and Chicana/o goodlife writing.

CHAPTER DESCRIPTIONS

This study begins in the second half of the nineteenth century to show how Mexican American culture before the U.S.-Mexican War sustained a cooperative and sustainable relationship with the natural environment in ways that were nearly destroyed during the transition to U.S. political, social, and cultural dominance in the newly acquired Mexican territories. Chapter 1, "Epistemological Hierarchy and the Environment: Erasure of Mexican American Knowledge in Three Nineteenth-Century Novels," looks at two works by María

Amparo Ruiz de Burton and one work by Jovita González. These writers depict the power transition in California and Texas, respectively. Ruiz de Burton's 1872 novel *Who Would Have Thought It?* satirizes the abolitionists in the North during the U.S. Civil War and critiques U.S. ignorance of Mexican American culture in the newly acquired West. Her 1885 novel *The Squatter and the Don* shows the rarely seen Mexican American point of view amid the struggles over land possession that followed the U.S.-Mexican War (1846–1848) in southern California. Along the same lines, Jovita González takes us into the domestic life of an hacienda during the U.S.-Mexican War, writing in hindsight from her 1930s South Texas perspective and offering an interpretation of nineteenth-century prejudices through the lens of her own experiences with Jim Crow segregation. Her novel *Caballero* shows us how one man's love for his land drives him to increasing degrees of madness when he believes he will lose his hacienda to Anglo-Americans, at once showing the Mexican American passion for land and the contradictory destruction that possessiveness can bring. This chapter sets out the original trauma that defines Mexican American goodlife writing—land loss—and anticipates how this experience leads to the eventual rejection of the idea of possession itself.

Chapter 2 takes up the writings of Mexican Americans who gained access to positions of privilege during the early twentieth century by means of claiming varying degrees of "whiteness," and in their writings reveal their turmoil in this process. This chapter, titled "The Coloniality of Being and the Land: Identity in Early Twentieth-Century Goodlife Writing," details how writers Fabiola Cabeza de Baca, Eva Wilbur Cruce, Nina Otero Warren, Jovita González, and Sabine Ulibarrí show the ways Mexican Americans lamented their loss of connection to the land in proportion to the degree to which they gained access to white privilege. These writers show the inner trauma caused when individuals choose not to delink from colonial/modern structures that alienate them from the land and from one another. Otero Warren, who claimed a Hispano identity, writes nostalgically about the Indian and the mestizo proximity to the natural environment, and also signals the ways in which she remains wistful regarding those times but not necessarily apt to trade her privilege for those close-to-nature experiences. Ulibarrí takes a different approach. His story "The Man Without a Name" concerns a man's gradual insanity at being distanced from his family's ancestral lands, which includes his father's voice haunting his every moment. Finally, Cabeza de Baca, Wilbur Cruce, and González use different genres—two memoirs and folktales, respectively—to show how claiming Mexican American mestizo identity offers the best approach to one's sense of self, access to the broadest cultural wisdom, and maintenance of a positive relationship with the land. Of the writers in this chapter, these three are the only ones who present narratives with evidence of delinking from colonial/modern structures. Until very recently, early twentieth-century Mexican American writings received little critical attention, perhaps precisely for the reasons this chapter finds them relevant and rewarding to examine: their preoccupation with racial hierarchy and privilege, as well as the psychic trauma they generate. The continuing influence of ethnic hierarchy has a lot to teach environmentalism, even today, so long after the Civil Rights Movement, and as we find ourselves in a new moment of ethnic backlash, with the power of civil rights landmarks such as the Voting Rights Act seriously curtailed as just one example.[18]

Chapter 3, "'La Santa Tierra': Chicana/o Writers Transcending Possession in the Late Twentieth Century," takes a look at the goodlife writings from the last half of the twentieth century, when Mexican Americans joined the Civil Rights Movement and changed their identity to Chicana/o in order to stand up against the elitism that oversimplified their ethnic identity and attempted to assimilate their culture. The Chicana/o Civil Rights Movement famously included protests as well as lawsuits to return lost lands to descendants of colonial land grant beneficiaries. Despite the popular understanding of this era's concern with land repossession, chapter 3 shows the decolonial transition to explicit expressions of disdain for possession. Specifically, Chicana feminist writers published essays and opinion columns in political newspapers that fundamentally challenged the desire for repossession of lands promoted by movement leaders such as Reies López Tijerina. This contests a major aspect of Chicana/o civil rights history. Informed by the desire to heal the trauma of dispossession and by their own experiences of domination due to their ethnicity and their gender, the Chicana feminists sought to respect nature's autonomy rather than further restrict it by enforcing ownership. I use political movement newspapers from the archive to support this claim, especially from the northern New Mexico newspaper *El Grito del Norte* (1968–1973) and the writings published there by Elizabeth "Betita" Martinez and Enriqueta Vasquez. I also include a reading of *Bless Me, Ultima* by Rudolfo Anaya that allows for what has heretofore been considered an anomalous appearance during the Civil

Rights Movement, given its lack of explicit political protest. Ultima's organic approaches to life and healing better fit into the goodlife writing genealogy.

Chapter 4, "Active Subjectivity in Migrant Farmworker Fiction: Rejecting Alienation from the Land," takes up Tomas Rivera's *And the Earth Did Not Devour Him*, Helena María Viramontes's *Under the Feet of Jesus*, and singer-songwriter Tish Hinojosa's folk tune "Something in the Rain." I read these works about the migrant farmworker experience to portray capital's attempt to alienate Mexican Americans from the land alongside the persistent goodlife communion with the natural environment, even in the harshest conditions. The decolonial concept of "active subjectivity," developed by theorist María Lugones, helps to reveal a farmworker subjectivity that is neither passive nor overtly activist. The works by Rivera and Viramontes have long been associated with a nascent activism, but I read these writings alongside the little-known song by Hinojosa to show a decolonial ability to critique oppression in a useful way that does not always mean one marches in the streets or runs for political office. Decoloniality requires the act of delinking from the dominant hierarchies of power, and these works show this subtle process in action in one of the most politically, socially, and economically vulnerable populations in the United States.

Finally, chapter 5, "Ecology and Chicana/o Cultural Nationalism: Humility Before Death in Cherríe Moraga's Millennial Writings," considers more recent writings that also make the most explicit references to environmental issues. Contemporary writer Cherríe Moraga directly challenges phenomena such as farmworker abuse, gentrification, and consumerism that combine to show the ways that the ongoing tradition of goodlife writing offers parallel and non-destructive practices. A discussion of three specific works by Moraga rounds out this study's literary history of Chicana/o environmental writing. *Heroes and Saints*, one of her early plays; *The Last Generation*, a strong collection of poetry and prose; and *The Hungry Woman: A Mexican Medea*, a haunting play set in a dystopic future, all put Chicana/o cultural production into conversation with the ecological circumstances of the late twentieth century. All three works betray a strong analysis of transnational dynamics and the global currents of politics and capital that put Chicanas/os and other peoples of color in the position of paying the world's ecological debt. *Heroes and Saints* is Moraga's most overtly ecological work, and her best-known theater piece. It is a play that reveals the injuries, literal and figurative, that toxic environments inflict on a small agricultural community in the central San Joaquin Valley of California. In her collection titled *The Last Generation*—a title that challenges future generations to continue the political struggles of their ancestors—Moraga delivers some of her most impassioned pleas on behalf of a reciprocal and sustenance-oriented human relationship with the environment. These poems and essays reveal a writer who cares deeply for both her culture and the natural environment that sustains it. *The Hungry Woman* (2001) reinvents key myths from both Western and indigenous traditions to show that legendary women throughout history sought to deconstruct hierarchy and exploitation. She invokes the past in order to instruct the present, and, in so doing, she educates the contemporary Chicana/o generation about the errors committed in the early movement, especially in regard to stringent nationalism. In these days of growing struggles to forge viable relationships between the local and the global, Moraga's consistent engagement with the theme of nationalism becomes particularly relevant. Indeed, nationalism has been a recurring theme for Chicana/o writing ever since the 1960s. Moraga makes sense of this nationalism in relationship with environmental issues, reinventing the Chicana/o relationship to territory.

CONCLUSION

Mexican American goodlife writing preserves an approach to the natural environment that has resisted and continues to work against the destructive practices that brought us to our current environmental crisis. Mexican American approaches survive at the margins and therefore make little impact on large-scale environmental policy and practices, but, importantly, they survived up to today, when we can perhaps begin to heed them again. Such hardy ideas deserve attention. While environmentalism sometimes looks to scientific research for ways to address environmental problems, we cannot solely rely on technological fixes for looming crises like climate change, water shortage, and toxic contamination. Even if scientists develop strategies, it still remains for humans to put them into practice. Everyday human culture must play its role and establish practices that sustain a healthy and livable environment. Sometimes this means participating in new practices, like the passive harnessing of energy by means of solar panels on rooftops. But we can also look to the past in order to access sustainable environmental cultures. In the Mexican American past we can find, for example, adobe homes that were built with nearby materials that blend in with the local landscape and work to keep indoor temperatures cool in the summer and warm in the winter.[19] By looking to the past, we can access intuitive practices that are already part of an established culture—so that it is less of a challenge to integrate into everyday lives. Practically speaking, it is to our advantage to implement both the new and the old practices to meet environmental challenges—a landscape of adobe homes topped with solar-paneled roofs.

Patricia Preciado Martin's short story "Earth to Earth," in the collection *Days of Plenty, Days of Want* (1999), shows how reading Chicana/o goodlife writing relies on an understanding of the decolonial. Martín's touching story about the lifespan of an adobe home in Tucson, Arizona, shows how the Mexican American family who built the home with love and a sense of intimacy and respect for the local landscape in the 1910s eventually loses the home due to the assault of progress and economic development in the 1970s. She makes her point plain with a stroke of irony: the city demolishes the adobe home in order to build a "Mexican style village" that will draw tourism to the city center (Martín, *Days of Plenty* 37). This story and many other works that this book discusses show how modernity and a history of colonization correlate to destroy traditional and organic approaches to the environment and exploits peoples of color, leaving them exposed to the harshest aspects of the natural environment—shelter loss, manual labor, toxic environments, and weather extremes. Yet peoples of color staunchly refuse to alienate themselves from nature, continuing to work with the natural environment as an ally rather than an enemy in need of domination.

At this point, my reading of Mexican American goodlife writing might appear to approach essentialism. Let me emphasize that what I include and assess here is a practice that does not depend upon an essential human nature of any one sort. To begin with, Mexican American identity builds upon its mixture, not an essentialized purity, and draws strength from its history of *mestizaje*. Also, this is a literature of great beauty and poignancy that records a long history of active resistance to epistemic assault, preserving and generating a depth of connection and cooperation between humans and nature while it also challenges readers to move beyond the modern/colonial paradigms and toward decolonized knowledges. Our current ecological crisis requires the radical proposition that we engage the ideas and processes that have not succumbed to modern and colonial ideals and values.

1

EPISTEMOLOGICAL HIERARCHY AND THE ENVIRONMENT

Erasure of Mexican American Knowledge in Three Nineteenth-Century Novels

THE U.S.-MEXICAN WAR AND THE ENVIRONMENT

MAJOR EVENTS OF AMERICAN ENVIRONMENTAL HISTORY include the establishment of the national park system, the Dust Bowl, the publication of Rachel Carson's *Silent Spring*, the establishment of Earth Day, and the emergence of environmental justice movements across the nation. The U.S.-Mexican War never figures into this story of American environmentalism. However, one cannot understand the Mexican American contributions to environmental thought without starting with this war, through which Mexico lost half its territory to the United States and when approximately one hundred thousand Mexicans, without moving an inch, suddenly found themselves residing within U.S. boundaries. The early literary history of Mexican American writing and the environment begins with two novelists who show the way Mexican Americans negotiated Manifest Destiny's impact on their relationship with the land. Although the Treaty of Guadalupe Hidalgo pledged to protect Mexican Americans as new U.S. citizens with rights and responsibilities, including property rights, Mexican Americans lost possession of lands they never meant to sell.[1] No longer being able to run their own farms and ranches meant that the environmental knowledge Mexican Americans had cultivated for generations could not influence the management of lands in the U.S. Southwest. Nineteenth-century Mexican Americans were prevented from implementing their goodlife practices, longstanding knowledge of aridity, irrigation, and the local dynamics of flora and fauna in the Southwest, and ever since, Mexican Americans scrutinize others' land management. Interestingly, the initial moments of Mexican American goodlife writings concern the impact that Manifest Destiny had on the domestic sphere rather than dwelling on land management issues. These two writers, who are Mexican American women, thus implicitly deconstruct the conventional gender divides of public/private : masculine/feminine at the same time that they reject the way that Americanization suppresses Mexican American contributions to shaping the land. Their works provide ample opportunity to perceive the way this historical watershed marks Mexican America's relation to the natural environment through the lens of land possession, dispossession, and the beginnings of transcending possession. This period is fundamental to understanding the unfolding of goodlife writing.

The two Mexican American writers to pen the best-known novels of the events during and following the U.S.-Mexican War are Jovita González and María Amparo Ruiz de Burton. Jovita González wrote during the 1930s but set her novel *Caballero*, coauthored with Margaret Eimer (pen name Eve Raleigh), during the very important three years of the U.S.-Mexican War (1846–1848). González did not benefit from seeing her novel published in her lifetime; scholars José Limón and María Cotera found the manuscript in an archive and published it in 1996. Her novel concerns the fate of a ranch family whose patriarch fights to keep Americans from taking over his lands. Readers benefit from all three eras that impact the creation of this novel—the 1930s when it was written, the mid-nineteenth century when it is set, and the late twentieth century when it was published. María Amparo Ruiz de Burton wrote and set her two novels *Who Would Have Thought It?* (1872) and *The Squatter and the Don* (1885) in the nineteenth century during the years following both the U.S.-Mexican War and the Civil War. *Who Would Have Thought It?* concerns the twin fates of a young Mexican girl and the Western territories the United States appropriated under the terms of the Treaty of Guadalupe Hidalgo, while *The Squatter and the Don* portrays the fate of another ranching family that struggles to keep its lands after the war. The way these two authors depict the plight of the land parallels the way they understand how landed Mexicans, newly become Americans, were fated to become dispossessed and marginalized during the new Anglo-American era.

These novels offer a critique of land dispossession and the missed opportunity to implement goodlife values and practices. But these novels take a broader view in their interpretation of this dynamic: they also show how the United States' lack of respect for the relationship that Mexican Americans forged with their ranchlands reinforces the same hierarchical discrimination that created the lack of regard for women and other disempowered populations, such as servants and veterans. Both novelists exhibit concern with the way that the U.S. legislative and court processes, as well as outright white-collar banditry, violated the land rights granted to Mexican Americans by the Treaty of Guadalupe Hidalgo. Moreover, both novelists, and especially González, also expose the destructiveness inherent in the notion of possessing nature and seeking to dominate the lives of others. These novels critique such objectification by putting women in the central roles as the agents of major influence and change, who implement values of dignity and respect in their decisions rather than attempting to exert power and control—values that begin to work toward transcending possession. Already at the beginning of goodlife writing, these authors show how these macro-scale politics and transactions also mark the micro-scale of the intimate and domestic.

Young, intelligent, and beautiful, albeit marginalized, Mexican women play the central roles in both González and Ruiz de Burton's novels. González depicts two daughters of a Mexican hacendado, Susanita and María de los Ángeles (or Angela), whose alienation from their father Santiago throughout the novel parallels the Mexican population's alienation from the dictator Santa Anna. With this analogy, the novel shows the way that the young nation reproduced the colonial legacies of patriarchy and racism even after independence from Spain. The novel argues that the continuation of these colonial hierarchical values left Mexico vulnerable to losing the U.S.-Mexican War as the exploited populations, such as servants and women, came to identify with U.S. ideals of equality and self-determination and eventually undermine the Mexican cause in favor of the U.S. takeover. Of course, history shows that the United States operated on patriarchal and racist hierarchies as much as Mexico did, but González presents an idealized version of the United States in this novel. This strategy challenges Jim Crow–era, as well as present-day, readers to reflect on the U.S. failure to live up to its promises of equality proffered to Mexicans during the U.S.-Mexican War and in the Treaty of Guadalupe Hidalgo. Indeed, González's critique of patriarchy and racism overrides any exception the novel may take to the

U.S. imperial takeover of half of Mexico's territory.[2] This novel looks to the war as a historical moment when respect for the influence that women and others in vulnerable positions had on the events of the time could have offered both nations an advantage at this important turning point.

Caballero shows the way González makes a strong case against patriarchy and racism—in the context of her novel, it is Mexican patriarchy and racism; in the context of the time and place in which she writes, it is American patriarchy and racism. González's novel introduces nineteenth-century conflicts with almost a century's worth of hindsight. Ruiz de Burton, on the other hand, offers a real-time nineteenth-century view of the years following 1848. Her young Mexican women, Dolores and Mercedes, demonstrate the ways that the United States took advantage of a vulnerable population and erased the Mexican American contribution to shaping the land we now call the U.S. Southwest. As their names, Dolores and Mercedes, suggest, the young women suffer cruel trials associated with land and its resources in the years following the U.S.-Mexican War. Dolores, of course, means "sorrows," and Mercedes can mean "mercy" and also "land grants." In this context, the secondary meaning of Mercedes's name becomes more significant because the Mexican American population gained its lands through Spanish- and Mexican-era land grants that were often invalidated in U.S. courts after 1848, transforming a ranching aristocracy into a laboring class in a few short years (Montejano 50–74; Griswold de Castillo 62–107).

González's ancestors were among those who lost lands, and she felt the consequences in her own life through her family's lack of economic resources. Still her parents made education a priority for their children, and only during her college years did González have to work to support herself. Ruiz de Burton was a young woman during this time of transition, so it is not surprising that she writes indignantly about the injustices that she witnessed and suffered first-hand (Aranda 95). Scholars recognize that hers was a voice of opposition to the U.S. political collusion with Anglo-American businessmen to exploit Mexican Americans and to take over their lands (Aranda 81–117; J. Rivera 82–109; Alemán). Scholars also agree that both González and Ruiz de Burton articulated feminist arguments in their novels, showing how men's blindness to the central role that women play in the events of all three novels occasions needless harm (Cotera, "Hombres Necios"; Sánchez and Pita; Casas 159–67). However, little attention has been given to the way that González's and Ruiz de Burton's writings lend a Mexican American sensibility regarding the natural environment.

These novels show how nineteenth-century Mexican Americans practiced coexistence with nature and suffered at the hands of American imposition of racist and imperialist agendas on the land and its peoples. This, I contend, represents a key contribution that Mexican Americans can make to environmental discourse and shows how Mexican Americans have a significant impact on U.S. environmental history. Far from celebrating the designation of lands as protected by state or federal oversight, these novelists offer an implicit critique of the idea of "wilderness." The West, to which Horace Greeley urged footloose young Anglo men to go, was not wild. It was well settled and well used by its Mexican as well as its indigenous inhabitants (Peña, *Mexican Americans*; Keller and Turek).

THE GOODLIFE'S DISSONANCE WITH WILDERNESS AND ANGLO-AMERICAN AGRICULTURE

These nineteenth-century novels were written and/or set in the early days of national parks, some of which—like the iconic Yosemite—the United States established in newly acquired Mexican territories. Indeed, nature writer and conservationist John Muir's descriptions of Yosemite convinced many U.S. politicians, including President Theodore Roosevelt, to approve the preservation of natural spaces, making Yosemite a benchmark in the development of wilderness parks (Fox 103–47). Environmental historian William Cronon notes, in his now well-known critique of the wilderness idea, "Yosemite was deeded by the United States government to the State of California in 1864 as the nation's first wild-land park, and Yellowstone became the first true national park in 1872" (Cronon 474). Yosemite became a national park in 1890. But before the United States could deed Yosemite to the State of California and then later make it an iconic national park, it had to start and win the U.S.-Mexican War to gain these lands. Neither environmental studies nor Chicana/o studies make note of this war as a key transition relevant to environmental history, yet the takeover of Mexican lands helped make possible one of the most important ideas that we have used in the twentieth century to inspire environmental sentiment: wilderness. To make the idea of wilderness work, the United States erased not only the American Indian but also the Mexican American transformation of the natural environment in California, Nevada, Utah, New Mexico, Texas, Arizona, and parts of Wyoming, Colorado, Utah, and Kansas—all former territories of Mexico.

Moreover, environmentalism's active promotion of wilderness has served to alienate people of color up to the present day (Guha, "Radical American Environmentalism"). Cronon notes that the "removal of Indians to create an 'uninhabited wilderness'—uninhabited as never before in the human history of the place—reminds us just how invented, just how constructed, the American wilderness really is" (482). In the case of Mexican Americans, wilderness and the national parks serve as a stark reminder of what González and Ruiz de Burton put at the center of their novels: that nineteenth-century American imperialism disenfranchised and dispossessed the Mexicans who chose to stay on their lands, lands suddenly located in the United States after 1848. Important for an environmental analysis of these three novels, the imperial attitude also kept the United States from gaining the significant skills, knowledge, and loyalty of the new Mexican Americans. This lost repository of knowledge and loyalty becomes ecologically significant when one considers the fact that landholding Mexicans had learned a great deal about how to sustainably manage their haciendas, what I call their goodlife values and practices that survive in the literature by Mexican Americans today.

U.S. imperial fervor obscured Mexican American environmental knowledge in favor of two broadly environmental goals that would develop over the later nineteenth century: ultimately, to create national parks in "wild" areas that would help establish a U.S. national identity based on iconic places; and in the short term, to use what Anglo-American farmers considered underdeveloped Mexican American ranchlands to implement tried and true agricultural practices—tried and true in the well-watered North American East and Midwest—that were unfortunately inefficient and unsuitable to the climate and terrain beyond the hundredth meridian. Concerning the goal of creating wilderness, Americans felt they had lived too long in the shadow of ancient European civilizations, and they looked to the newly acquired West and its monumental scenery to compete. Historian Alfred Runte says of this period, "The agelessness of monumental scenery instead of the past accomplishments of Western Civilization was to become the visible symbol of continuity and stability in the new nation" (12). Ruiz de Burton's Dolores in *Who Would Have Thought It?* exposes U.S. hypocrisy, especially through violations of the Treaty of Guadalupe Hidalgo, concerning the lands and peoples it abused in order to provide these national symbols.

As to the latter goal concerning competing land use approaches, Mike Davis has observed that, in contrast to the American Indian cultures, as well as the Spanish and Mexican colonizers, the "Anglo-American conquistadors . . . were riven by confusion and ambivalence" concerning the semiarid climate of the newly acquired territories (224): "Immigrants from the humid states, moreover, brought with them deeply engrained prejudices about climate and landscape that had been shaped by the environmental continuum of the Northwestern Europe and the Eastern United States" (225–26).

Ruiz de Burton's *The Squatter and the Don* shows how Mexican American ranchers attempted to educate the newly arrived eastern Americans about the unique challenges of semiarid agriculture, but California state laws quickly stepped in to support an imperial agenda that discouraged Anglo-Americans from working with Mexican Americans.

All three novels reflect on the failed potential alliances between Mexicans and Americans in the latter half of the nineteenth century by focusing on Mexican American women. *Caballero* challenges its readers to imagine the way the war's impact could have been different, if Mexican women, who would have fostered cooperation between Mexicans and Americans, had been empowered. In the pages of Ruiz de Burton, it is no coincidence that Dolores travels to the Northeast with a treasure trove of gems from the Western United States in *Who Would Have Thought It?* and that *The Squatter and the Don* begins with a debate between a Mexican rancher (the don) and a group of Anglo squatters—one of whom will eventually court the don's daughter Mercedes. With her two female protagonists, Ruiz de Burton shows how the U.S.-Mexican War created the possibility for Mexicans to become new and empowered U.S. citizens and for Americans to work together with them for a common good: a sound democracy and development of beautiful, fertile, and productive territories. However, U.S. greed for the lands and mineral rights of Mexican American citizens resulted in an alienation of the new citizens from their lands as well as from their new nation.

We live with the legacy of this forsaken cooperation to this day. Recent studies indicate that Latinos are the ethnic group most exposed to toxic harms. Secretary of Labor Hilda Solis (in office 2009–2013) states in a foreword to a Natural Resources Defense Council study on urban toxicity, "Our [Latino] children disproportionately suffer from asthma as a result of air pollution, and pregnant women and the elderly suffer from lead and other contaminants in our drinking water" (Solis iv). A University of California–Irvine research team has shown how there is presently a "disproportionate proximity to toxic waste among Latinos" in Southern California (Hipp and Lakon 681). Moreover, the semiarid climate of Southern California continues to be misunderstood by its current population (Davis). Both González and Ruiz de Burton recount the pain of the initial steps toward environmental degradation and social injustice in their three artfully crafted novels, written in a unique blend of sentimental Victorian romance and matter-of-fact social criticism.

CABALLERO: THE UNITED STATES' AND MEXICO'S HYPOCRISIES, FALSE HOPES

Caballero makes use of its setting during the U.S.-Mexican War not to present scenes of battle and bloodshed but to show what happened away from the battlefields, in the courtyards, drawing rooms, churches, bars, and ballrooms of a traditional ranch—Rancho La Palma—and the small Mexican community of Matamoros. This novel, written in English, gives readers a rarely glimpsed view of the war from a domestic Mexican perspective, importantly including the experiences of the women, children, clergy, and servants of the embattled republic. The novel also offers the intimate thoughts of the rancho's patriarch, don Santiago de Mendoza y Soria. The polar opposition between the wishes of don Santiago and the actions of his family, particularly daughters Susanita and María Angela, and his servants reflects the Mexican republic's susceptibility to U.S. influence during the middle of the nineteenth century. Santiago attempts to wield absolute control over his family and hacienda just as President Santa Anna of Mexico tried to wrestle the nation into a dictatorship, leaving Mexico vulnerable to U.S. subversion. However, rather than responding to the United States in the same way their father did, Susanita and María Angela take approaches that allow for love, generosity, and dignity. Far from portraying the Mexican people as victims of the United States, *Caballero* challenges its readers to consider the idea that Mexico fell prey to its own patriarchal, racist, and dictatorial regimes, and that perhaps the present-day oppression of ethnic minorities and women in the United States result from the same sources of patriarchy, racism, and the allure of power. In the idealized ending of *Caballero*, Mexicans (in particular Mexican women) willing to compromise with the Americans enjoy a happy life in the new nation. Historical reality, however, tells a different story. The idealized ending prompts readers to note the dashed hopes of the new Mexican Americans.

The U.S.-Mexican War was the third major shock of the nineteenth century for Mexicans. First, the republic fought from 1810 to 1821 to earn its independence from Spain. In 1835 Mexico fought to keep Texas from rebel American immigrants whom Mexico had allowed to settle the territory, which was promptly renamed the Republic of Texas in 1836. Once the United States accepted Texas as a state, the dispute over whether the Nueces River or the Rio Grande formed the southern boundary led U.S. troops to invade Mexico in 1846. With communication difficult over sparsely populated lands, the Mexican population within the disputed swath of land between the Nueces and Rio Grande found themselves in the middle of a war zone with little access to information regarding their fate. This is where *Caballero* begins, with a neighbor bringing news of the war to the Rancho La Palma while its patriarch, don Santiago, oversees the morning prayers.

What follows is a novel written as a combination of the eighteenth-century sentimental novel and the nineteenth-century domestic novel, complete with star-crossed lovers, blushing cheeks, and fainting maidens as well as gallant horsemen and loyal servants. The novel uses these formulas to offer incisive social critique as it allows the women a critical voice and the peons an escape from cruel masters. In this sense, *Caballero* uses its setting in the nineteenth century to present a familiar story, but it empowers its characters in a way more akin to the twentieth century. Still the shifts work in the way the novel addresses a more contemporary audience, challenging readers to rethink their own era's efforts to control women, people of color, and even nature itself.

Four episodes in particular give this novel its place in Mexican American goodlife writing history. The first occurs as the Mendoza y Soria clan reflects on its predicament of inhabiting disputed territory; they have no idea under whose flag they are living. Don Santiago attempts to decide the fate of his clan, while the rest of the novel shows his power diminishing over time. The way he responds to this situation—by trying to wrest control by any means necessary—offers a negative example of how to respond to this situation. He emerges as a goodlife antihero. The second involves the way the women in the family, particularly the widowed sister of Santiago, doña Dolores, and Santiago's wife, María Petronilla, successfully negotiate tense moments with Americans by opting for dignity rather than choosing to display authority or exert control. They emerge as goodlife heroes. The third concerns Santiago's unrealistic effort to redirect the flow of the Rio Grande in a futile attempt to keep his lands under Mexican rule. The fourth reveals the novel's ultimate judgment regarding don Santiago's authority by showing readers how he dies. Both of these last two episodes show how Santiago represents the goodlife antihero.

As the Mendoza y Soria clan faces the predicament of not knowing whether they live under Mexican or American governance, several options seem clear to don Santiago, the patriarch: fight against, manipulate, or compromise with the Americans, the less familiar rulers. Don Santiago considers these options while leaning with his back against the palm tree planted by his grandfather when he established Rancho La Palma de Dios in 1748. His grandfather had been a royally appointed surveyor for the Crown, and he traveled the territories with an eye for finding a place of his own in the northern reaches of New Spain's territories in the New World. When he saw this swath of land south of the Nueces, "thirty miles north of the Rio Grande and some sixty miles west of the Gulf of Mexico," the novel describes it as a place that "beckoned and smiled" (xxxvii, xxxvi). On the first page of the novel's foreword, nature already plays an active role, seemingly seducing the colonizer to settle there. This moment offers an ironic commentary on how patriarchs such as Santiago and his grandfather José Ramón feel that the land chooses them to dominate and control it. In other moments, Santiago takes an almost maniacal tone while enjoying his possession of the hacienda and all its inhabitants, human and nonhuman alike:

> "Mine," he murmured. "All this that I can see and far beyond, is mine and only mine." . . . Power was wine in his veins. Power was a figure that touched him, and pointed, and whispered. Those dots on the plain, cattle, sheep, horses, were his to kill or let live. The *peons*, down there, were his to discipline at any time with the lash, to punish by death if he chose. His wife, his sister, sons, and daughters bowed to his wishes and came or went as he decreed. (33)

So strong is his impression of total control that he perceives power as "a figure that touched him" and assures him that he determines whether anyone on his land lives or dies. Despite his certainty of possession over the hacienda and its inhabitants, don Santiago eventually experiences an almost complete subversion of his authority, primarily due to the influence of Americans and their lack of interest in exerting such power over women, servants, and the land. This is, of course, an ironic critique of Mexicans by way of Americans because when González and Eimer are writing this novel, they are doing so under the constraints of Jim Crow segregation as their everyday experience.

Almost one hundred years after his grandfather José Ramón establishes Rancho La Palma de Dios, Santiago leans against his grandfather's palm tree and debates with himself about whether or not to compromise with the Americans. Ultimately he decides: "There would be no compromise, neither for himself nor for those over whom he was master . . . He would have to hate the *Americanos*, for they were evil" (21). As he makes this decision, he feels the palm tree "strong against his back, like a firm pillar, for it was a symbol" (21). Santiago takes it as a symbol of the ranch that he must protect, but the novel suggests it draws its symbolism from its biblical context as the palm branches of Christ's Palm Sunday, which stand for peace and cooperation: "The leaves of the palm rustled softly, the night held him gently. There was no one there to tell him that hatred feeds upon itself and lives but to destroy. No one to remind him that the Christ who rode in a procession of palms came to earth to preach the gospel of love" (21).

No person was there to "remind him," but the palm itself attempts to exert a peaceful influence on his decision, "rustling softly," and showing the natural environment's agency by promoting "the gospel of love." Santiago's daughters, who respond to the courtship of two American men, heed the transformative power of love, unknowingly following the palm tree's initial suggestion. Susanita, upon reflecting on her belief that God disregards differences in his omniscient love for all, also recognizes nature's agency: "Looking up she caught the twinkle of the first evening star, pale silver against a sky still blue, and nodded to it as one greets a friend. That was part of God, that star and it was there for everyone to see and love" (10). Santiago, on the other hand, ignores the power of peace and love. Of course, it is also ironic that these suggestions of peace and love occur in the context of Christianity, which has its own history of collusion with colonization to destroy cultures and dominate indigenous peoples.

Don Santiago's widowed sister and his wife, however, heed the values of dignity and respect in their dealings with the Americanos. In so doing, both women foster the relationships between young Susanita and María Angela with their American suitors, and thus help the family successfully transition to the era of U.S. governance. Rather than serving as a manual for assimilation, *Caballero* shows the women mindfully negotiating a changing situation in order to ensure their own and their culture's survival. For example, doña Dolores displays due dignity during a crisis when all the hacienda families travel to the nearest city in Matamoros, Mexico, for the winter season to celebrate holidays and attend the traditional Catholic masses and celebratory balls. The occasion also serves to bring together all the eligible young people to help determine marriage matches.

An unexpected guest appears at the first ball, held at the Governor's Palace and known as "the most brilliant of the pre-Christmas dances" (89), and disrupts the expected pairings of the season. The arrival of the influential Mendoza y Sorias is much anticipated, of course, and upon their fashionably late entrance around midnight, the company of each daughter is graciously requested by proper young men acquainted with the family since infancy saying, "May I, Don Santiago, . . . have the honour of dancing with Susanita?" and "May I . . . be honoured by Angela's company?" (89). These were the courtesies taken for granted at such gatherings for generations, but the ball of 1846 sees a great departure from convention with the arrival of a gentleman from Virginia—Lieutenant Robert Warrener of the U.S. Army. Warrener and Susanita already know one another by sight from a chance encounter that sparked feelings of love between them, and it only seems natural to them that they dance together at the ball. However, in the eyes of the hacienda families, Warrener's mere presence at the ball was an affront: they suffered "shock at seeing a member of what was considered an inferior people" (93). When Warrener takes Susanita into his arms without asking permission and not even with the benefit of an introduction, the entire festivity collapses and the star-crossed lovers do not even notice that they are the only couple left on the dance floor after "scandalized *mamás* and outraged *papás* plucked their daughters from the dancing space where embarrassed escorts had been too bewildered to lead them to their seats" (93–94). The only thing that prevents all-out violence from ensuing is the code of hospitality to which the hacienda families adhere. Alvaro, the eldest of the Mendoza y Soria siblings and don Santiago's heir, gives Warrener a sharp slap in the face but restrains himself from more. Don Santiago responds to Warrener's insult by deciding to return to Rancho La Palma at first light in order to escape the humiliation of facing his peers after his daughter so willingly welcomes an Americano.

The rest of the Mendoza y Soria family, however, prefers to stay in town and continue with the holiday festivities. Susanita especially has an interest in staying close to the man she now feels confident she loves, Warrener. But how can they break don Santiago's seemingly ironclad resolve to leave right away? His sister, doña Dolores, shows don Santiago that the path of dignity and respect, goodlife values, is far preferable to that of outrage at a perceived injustice. When don Santiago encounters the other hacendados in the plaza and they confront him with his daughter's unbridled behavior, he smugly observes that his daughter simply respected the mayor's invited guest with her "good manners" which "are the attributes of a gentleman and his family, quite superseding personal feeling" (99). With this pronouncement, don Santiago transcends, via doña Dolores's suggestion, the pettiness inherent in the idea that a man controls every last decision of the women in his family.

Unfortunately, he does not heed doña Dolores's lesson for long. When Warrener visits Rancho La Palma to formally ask for Susanita's hand in marriage, "Don Santiago was swept away from his moorings of courtesy" and he attacks Warrener with a cane, striking him twice on the head and drawing blood (230). When Warrener asks the specifics of don Santiago's objections to him as a son-in-law, don Santiago lists among many things that Warrener "took advantage of a weak and frightened woman to be put [under] his roof" (231). Indeed, Warrener encounters María Petronilla upon his arrival as he enters the patio at Rancho La Palma, and as hostess she must decide whether to welcome him or turn him away. Warrener quickly proffers his nascent Spanish to appreciate the shade offered by their lovely patio, and the conversation quickly turns to a comparison of domestic customs between Mexican haciendas and Virginian plantations. Warrener's gentlemanliness soon wins María Petronilla over and she invites him to meet with don Santiago, which leads to the unseemly violent reception described above. In this instance, only María Petronilla succeeds in abiding by dignity and respect for Susanita's autonomy, while don Santiago behaves as if Warrener threatens to steal his property. After suffering more trials, Susanita eventually marries Warrener, despite her father's objections, revealing don Santiago's efforts at controlling everything to be an obvious failure.

The significance of the goodlife values of dignity and respect in this social context—rather than in relation with the land—might not appear too clearly in the above example. After all, the dignity and respect with which doña Dolores and María Petronilla undermine don Santiago still emerge from aristocratic codes of conduct that harken back to the peninsula and invoke colonial hierarchies of race. But this history and cultural origin is exactly what makes these values' usage in this context compelling and important: Dolores and Petronilla keep the essence of the cultural values alive at the same time that they undermine the hierarchical structures these values originally meant to protect. In keeping the family in Matamoros, rather than returning to the rancho as Santiago wanted, Dolores facilitates the romance between Susanita and Warrener. By welcoming Warrener to Rancho La Palma with hospitality, Petronilla allows the two lovers one step closer to their dream of union, despite Santiago's extreme response. Even if the goodlife values of simplicity and sustenance are not in full display in this novel—the family lives lavishly, while the *peones* (laborers) live austerely—the goodlife values of dignity and respect help to expose the hypocrisy of a man's attempt to control his family, women, and the land. He cannot supersede the forces working against him from the outside as well as from within his own domestic sphere. Ultimately, he is the author of his own destruction.

After the triumph of the Americans becomes clear, Santiago even attempts to exert his will over the Rio Grande. He strikes upon a plan to make his land Mexican again: change the course of the Rio Grande so that his lands lie to the south of the border. He commands his vaquero: "Tomás, tomorrow you bring *peones* and dam the river so it flows way around" (304). His son Alvaro points out a flaw in his plan, "But, *papá*, that cannot be done. There would have to be a dam as high as a house and stronger than one to make the river take an elbow" (305). And his vaquero warns, "It is against the law of God to change what He has made" (305). Santiago fails in this endeavor, of course, and the novel shows how the natural environment retains ultimate autonomy.

Finally, the novel ends with Santiago lying dead on his land, holding a fistful of dirt, when his despised son-in-law finds him: "Warrener noticed that Don Santiago's hand was tightly clenched. He stooped and opened the fingers to see what they held. A scoop of earth, brown and dry, trickled from the palm and lost itself in the sandstones" (337).

Warrener, Susanita's American husband, allows the land to escape his clutches, symbolically showing at least the possibility for open-hearted Americans to be better land managers than bitter-spirited ranchers like Santiago, who presumably died from heartache over losing control over his land and his family. Santiago cannot hold onto the land forever, and when Warrener gets his opportunity to grasp it, he calmly lets it flow back onto the sandstones from where it originated. González and Eimer were signaling an ideal that turned out to be far from the reality.

This environmental reading of *Caballero* sheds new insight on scenes previously considered only within the context of the U.S.-Mexican political relations. *Caballero* makes clear how obsession over land possession drives the demise of one Mexican American patriarch. The novel also shows how an author like González exposes the parallel failures resulting from attempting to control nature as well as attempting to control women and servants.

WHO WOULD HAVE THOUGHT IT?: MEXICAN LANDS AND PEOPLES IN NINETEENTH-CENTURY AMERICA

Ruiz de Burton's *Who Would Have Thought It?* also makes a comparison between the fate of land and the fate of a woman. In the opening scenes of *Who Would Have Thought It?* the patriarch of an Anglo New York family, Dr. Norval, returns from a four-year trip out West. The novel is set in the years leading up to and during the U.S. Civil War. Dr. Norval planned his trips to the West as often as possible in order to pursue his amateur science of collecting rock specimens, and in the hopes of perhaps discovering a profitable location for mining. His wife, three daughters, and spinster sister-in-law await his arrival with mixed emotions. They look forward to the long-anticipated reunion, but dread the intrusion of the geologic samples that Dr. Norval customarily brings back from these trips. Far from depicting a scene in which wife, daughters, and sister-in-law clamor to receive gifts and tidings from an exotic place, Ruiz de Burton shows these women to be toughened against such expectations. Judging from the family's disdain, Dr. Norval's samples apparently never before amounted to anything more than burdensome curiosities. This, however, proves to be an altogether different collection of samples, beginning with the dark little girl at his side. Unexpectedly, the "rocks and pebbles" eventually transform the Norval women into wealthy socialites, and the dark little girl becomes the reason for their shift in fortune. Nonetheless, all but one of the Norval women choose to despise the little girl, even after she is no longer dark—her skin tone results from a dye used to color her skin while she was the captive of an American Indian tribe in the West. When she grows up, Dolores Medina, "Lola," becomes beautiful, light-skinned, wealthy, and well bred, but only after years of loneliness in the Norval household. This novel belongs in the goodlife writing history in the way it shows how the United States responded to Mexican lands and peoples in a way that simply looked to their commodifiable and exploitable traits and refused to see them in their full complexity as members of a human and environmental community. In young Lola, we see a Mexican character with the capacity to defy this imposed interpretation of her significance.

Lola's story is about as unbelievable as the idea that one nation would take over half its neighbor's territory and then "discover" a wealth of precious minerals and beautiful landscapes in it. In other words, this historical novel's seemingly far-fetched events suit the times in which they occur. Lola's mother is captured by "Indians"—the novel never specifies which tribe—from her wealthy family's hacienda when she is pregnant with Lola. Lola is born five months after her mother is captured, therefore proving her a legitimate child of her parents' union as well as attesting to her "pure Spanish blood," which was of utmost importance for affluent Mexican families of the time. The chief wants Lola's mother, Theresa, for his wife, but Theresa only agrees when she gets assurances from the Indians that no one will touch her daughter. When Lola is ten, the Indian chief turns ill and runs across Dr. Norval's explorer party. Dr. Norval agrees to treat the chief and subsequently meets Theresa. Theresa asks Dr. Norval to promise to take Lola with him, away from the "savages," when he returns to New York. Dr. Norval agrees to take Lola with him and to do all he can to reunite the girl with her long-lost family in Mexico. Anticipating Lola's departure, Theresa's health begins to decline. However, before she loses all strength, she manages to collect and deliver to Dr. Norval a cache of rough gold nuggets and raw precious gems. She collected the cache during the tribe's wanderings and saved it to assist in Lola's escape, if such a time ever arrived.

Lola and the rough stones share a fate: upon their arrival in New York, both are belittled and disdained, yet, upon closer examination, the "little girl very black indeed" becomes white, and the "rocks and pebbles" (13) that threaten to "scratch the oilcloth, which was nearly new" become precious stones and metals. Accordingly, the economic situation of the family improves, but their morality becomes corrupted as they try to separate the *compañeras*—Lola and her stones. When they succeed in separating Lola from her riches, they discard her without a care, but remain vigilant of her precious "rocks and pebbles." This plot mimics the Americans' attempts to cheat Mexican Americans and indigenous peoples out of their lands, especially once they learn that the lands contain precious minerals.

The only thing between Lola and utter powerlessness and anonymity is a document her mother dictates in the final hours of her captivity to a scribe and translator provided by Dr. Norval. Dr. Norval, in all honesty and honorable intention, plans to rely upon this document to help him locate Lola's father in Mexico. But because he does not speak Spanish, he has to depend on his associate to take down Theresa's story, including the important names of Lola's family. The associate then translates the document and mails it to Dr. Norval. However, the document never arrives at the Norval home, ending up at the dead letters room at a post office in Washington, DC. Without the document that tells him whom to contact about Lola's whereabouts, Dr. Norval must take charge of Lola's care and education as well as her gold and precious gems.

Two significant parallels develop here between Lola's story and the story of Mexican Americans and their lands post-1848. First, the people's fate and that of their lands relies upon a single document: in Lola's case it was her mother's narrative; in the case of the Mexican Americans it was the Treaty of Guadalupe Hidalgo, which ended the war and included a pledge from the United States to protect the rights and lands of Mexican Americans. Second, because of the U.S. government's careless treatment of the respective important documents (loss of the narrative and violation of the Treaty of Guadalupe Hidalgo), Lola and her precious stones, and the Mexican Americans and their valuable lands, become subject to the whims of the U.S. government, state and local officials, corrupt businessmen, and a cruel social hierarchy. Lola's trajectory in *Who Would Have Thought It?* mirrors the vexed fate of Mexican Americans in the nineteenth century. Ruiz de Burton sees herself—as do all her fellow upper-class Mexican Americans—as racially "white," tracing her lineage back to Spain, and as a member of a privileged and distinguished social class. Yet she also recognizes that the bulk of Americans in the wake of the U.S.-Mexican War view Mexicans as racially impure, putting them in a class little deserving of distinction or possessions (Menchaca 2011). In her novel, she attempts to show how this perception of Mexican American racial "darkness" is only a temporary effect, like the dye that darkens Lola's skin. The discovery of gold in formerly Mexican territories, such as California and Arizona, and the establishment of wilderness areas and national parks in former Mexican territories only serve to exacerbate the instinct Americans already had to disdain the newest Americans: Mexicans. Likewise, Lola's precious metals and stones make most of the Norvals disdain her. In short, Ruiz de Burton offers Lola's tale as a parallel to the cruel reality that many Mexican Americans experienced post-1848.

Despite her portrait of loss and betrayal, Ruiz de Burton nonetheless writes her narrative in a way that challenges Americans to see the ways Mexican Americans can still make a positive impact on their new nation. For example, on the night of Lola's arrival in the Norval home, Dr. Norval attempts to coax his wife to view Lola in a positive light: "'I think that Lola, instead of being a *burden* to us, will be a great acquisition. Don't you think so?' said the doctor, after his wife had toyed with the gold for some time" (25). In this intimate scene between husband

and wife, Dr. Norval struggles to convince his harsh wife to welcome the little girl to their home and attempts to manipulate her by means of her greed for gold. Mrs. Norval, for her own part, initially questions her first response to reject Lola:

> The despised black child she now would give worlds to keep. She would go on her knees to serve her, as her servant, her slave, rather than let her go . . . Thus ran the thoughts of the high principled matron. But never once did it occur to her that she had sent the child to sleep with the cook and the chambermaid, and she did not know that the little girl was now crying as if her heart were breaking, calling her mother between sobs . . . (30)

Here we see Mrs. Norval begin to reason that the little girl can bring distinction to her home and family with her riches. At this point in the story, it is still possible to associate Lola closely with her accompanying goods. If only the Norval family continued to think in these terms, Lola would not suffer a childhood and youth of loneliness and disdain. The Norvals could work with Lola to make the most of their fortunes together.

However, Mrs. Norval soon reasons that she can contrive to take the gold and gems from Lola for the Norvals' own profit and even rationalizes that her family deserves them as reward for allowing Lola to set foot in their home. She asks her husband, "Didn't her mother give you anything for taking charge of her daughter for life?" (29). Later in the novel, Mrs. Norval further attempts to rationalize her family's use of the riches that Lola does not deserve or even supposedly know how to put to good use, and her choice of epithet reveals her increasing hatred toward Lola in specifically racist terms: "And would that little nigger be so rich, and her girls so poor? Their new carriages and splendid horses and handsome house, after all, did not make Mrs. N. happy" (49). At the same time, Ruiz de Burton editorializes that the family's use of Lola's riches does not increase their happiness: it only makes them want more and care for Lola even less. The way Americans rejected Mexican Americans in the nineteenth century certainly gave them access to more lands, minerals, and freedom to establish national parks, but they lost the opportunity to establish alliances with deserving peoples.

History does not show Mexican Americans escaping a measure of discrimination and inequality until well into the twentieth century, but Ruiz de Burton gives Lola a hopeful future after years of suffering in the Norval clan. Dr. Norval continues to be her protector and confidant, but his increasing frustration with his family's greed and the hypocrisy evident in many sectors of Northern society during the Civil War drives him to take a risky expedition to the African continent. He entrusts Lola to the guardianship of his son Julian, who is beginning to fall in love with Lola but who is also dedicated to his service as an officer in the Union army, which often keeps him from Lola's company and confidences. This leaves Lola relatively vulnerable to the manipulations of Mrs. Norval and her secret lover, the one-time preacher turned military officer, John Hackwell. Lola narrowly manages to escape plots to completely deprive her of her property and self-determination only after Mrs. Norval's younger brother Isaac, of all people, discovers the lost letter containing Theresa's narrative about Lola's background and family ties. Isaac finds Lola's father and grandfather in Mexico and brings them to New York, where Lola lives with the Norvals. Julian, Dr. Norval, and a few other allies work to foil the plots against Lola, and she eventually returns home to Mexico and marries Julian.

The recurring refrain in the novel does not concern social discrimination as much as it critiques government failure to honor its commitments. Indeed, Ruiz de Burton critiques the Mexican government as much as she chastises the U.S. Congress. The narrator comments, "If Mexico were well governed, if her frontiers were well protected, the fate of Doña Theresa would have been next to an impossibility . . . If we were to trace our troubles to their veritable source, we would often reach, more or less directly, their origin in *our lawgivers*" (201). Of course, part of what weakens the Mexican government's vigilance over its borders is the U.S. impulse to satisfy Manifest Destiny and expand its own boundaries. The United States and especially the Northern abolitionists are subject to constant caricature in this novel, from the bumbling fool who becomes a hero of the Civil War because his horse is spooked to charge into battle, to Mrs. Norval who crusades on behalf of the enslaved, but indulges deep hatred toward Lola. For Ruiz de Burton, the U.S. legislators in their greed and collusion with Anglo-American businessmen kept the United States from welcoming Mexican Americans, from learning their goodlife values and practices. Such is also the case in *The Squatter and the Don*, where Ruiz de Burton further argues that Mexican Americans had vital land management skills and a thriving culture to offer the United States.

THE SQUATTER AND THE DON: "THE SINS OF OUR LEGISLATORS!"

The Alamar family in Ruiz de Burton's 1885 novel, *The Squatter and the Don*, owns an expansive cattle hacienda in the San Diego area. In this novel, the Alamares, especially the patriarch don Mariano Alamar, make clear their knowledge of California ecology. The title of the novel itself reveals a story of California land use history in the late nineteenth century: the Anglo-American squatters came to California from the East Coast of the United States in search of lands to settle and farm, while "the dons," or Mexican American hacienda/ranch owners, already had land-use practices well developed in California. In *The Squatter and the Don*, the proud Ruiz de Burton, who was born in 1832 into a family who owned lands in Baja California, depicts her compatriots' experiences. Ruiz de Burton contrasts the Mexican hacienda owners' land use to that of the new population of Anglo-American squatters. The point is not to romanticize *californio* (Spanish or Mexican Californian) ecological knowledge; history clearly reveals Spanish introduction of invasive species and manipulative technologies. However, the californios had acquired an intimate knowledge of the land they had worked and shaped for at least a century, and they could not help but object to the unwise agricultural practices of the newly arrived Americans. Of the three novels in this chapter, *The Squatter and the Don* most explicitly exemplifies the earliest literary history of goodlife writing, especially in the ways the Alamar family exemplifies sustenance, dignity, and respect.

Ruiz de Burton demonstrates californio knowledge through practice in regard to the land, particularly in an early scene. Don Mariano Alamar proposes a compromise to the Anglos who threaten his and his family's livelihood: they are squatting on his land and killing his cattle. The events in this novel offer a fictional depiction of similar events that occurred all over California in the late nineteenth century. A couple of factors contributed to the encroachment of farmers on California ranches—in the novel and in historical fact. Squatters settled on ranches that they erroneously believed available for new American settlers on the strength of the recent U.S. takeover. In actuality, the ranches had long been settled by Mexican californios and their agricultural practices included fruit orchards and vineyards, but their main focus was cattle. However, the squatters killed the californios' cattle because they claimed the cattle ruined their newly planted crops by wandering among the plants and eating the green shoots. In this instance, killing cattle—an activity sanctioned by the new American "no fence law"—was not just an act of crop defense. The "no fence law" passed by the California legislature in 1874 made the fencing-in of cattle the responsibility of the ranchers (the Mexican Americans), and it protected any "farmer" (the Anglo-Americans, who were often squatters) who shot cattle that wandered into their crops (Raup 65). Of course, the Mexican American ranchers never used fences prior to this law, and it would be a huge expense to erect them. The practice of Anglo-American farmers shooting cattle reduced Mexican Americans' ranch capital and made ranch owners less and less financially capable of defending their lands in the courts, giving Anglo-American squatters the legal and economic advantage.

In response to the above cattle-killing scenario, don Mariano calls a meeting to propose a compromise to the squatters. Although he could justifiably consider the squatters greedy invaders, the novel shows him attempting to find common ground by invoking the enterprise in which they are all engaged. He says: "The reason why you have taken up land here is because you want homes. You want to make money. Isn't that the reason? Money! money!" (Ruiz de Burton, *The Squatter* 86). The squatters reply in the affirmative to this proclamation, some of them even laughing. Then the don presents a plan to them by which they can stay on the rancho without paying for the lands they have settled, make more money, and (in his interest) protect his cattle.

In summary, the plan involves an agricultural shift away from the farmers' struggle to cultivate wheat and other grains in arid Southern California. He suggests that they instead cultivate the more bioregionally appropriate growth of fruit trees and vineyards, after an initial period of raising cattle. He says, "Plant vineyards, olives, figs, oranges; make wines and oil and raisins; export olives and dried and canned fruits" (88). This agricultural shift would enable two key changes for the don that would also benefit the farmers. First, the farmers would use less land—as orchards and vineyards require less space than grain production—and therefore it would be feasible for them to build fences to keep out the cattle. If the cattle were no longer threatening the farmers' harvests, they no longer would have any right or need to kill the cattle. They would have one less chore on their busy farms. Second, the don would gain allies in business rather than legal foes, and the farmers would benefit from the don's local environmental knowledge in order to work toward successful enterprise in the greater San Diego region. In short, the farmers would continue the activity for which they moved west: the cultivation of produce. They would simply change from one kind of cultivation to another more sustainable one, and they would shed the task of shooting helpless cattle every evening.

However, several details, left unattended, would surely impede the plan. First of all, the cultivation of successful fruit orchards and vineyards takes years. The farmers would have to find another enterprise to keep them in the black until the orchards and vineyards could yield. The don answers this query with a practical proposal: he would give each farmer a number of cattle, and give them each five years or so to repay him, at no interest. This meant that as long as they bred the cattle adequately, the reproduction of the cattle themselves would generate enough animals to repay the don. This question of repayment, of course, suggests that the don may want the farmers to pay for the lands they were currently squatting. However, the don assures them that they will all retain their homesteads, at no cost. He promises to stop fighting their claims in the courts and present each of them with a quit-claim deed.

Still, in the face of all these advantages, the squatters reject don Mariano's proposal. They scorn his plan on the basis of racial prejudice, saying "perhaps you understand *vaquering* [cowboying]; we don't," at the same time mocking the Spanish language. Unfortunately, in one of the book's greatest ironies, don Mariano's plan fails in part because the Americans cannot imagine themselves as vaqueros, or cowboys—an image that these days stands as one of the most globally recognizable symbols of the American West. By invoking their common enterprise, agriculture, and environmental knowledge that can help them work toward their economic goals, don Mariano hopes to strike a working compromise with the squatters. And this compromise, this bridging of cultures, could be possible, if it were not for the existence of a "higher law" the squatters could invoke in their favor. The law they invoke is not of any environmental or moral origin, but the "no fence law" handed down to them from legislators who sought to promote a familiar Eastern American agriculture that was out of place in semiarid California. Rejecting the don's proposal that they become ranchers, the squatters also reject the one standard that the don would hold them to if they insisted on planting wheat: to fence their crops. A squatter named Mathews explains: "I am not so big a fool as to spend money in fences. The 'no fence law' is better than all the best fences" (90). More than just protecting crops, the "no fence law" protects an imperial agenda, giving familiar American enterprise an upper hand at the cost of californio culture and environmental knowledge. While the squatters may be benefiting from the laws, the true culprits in the novel are legislators and monopoly capitalists, who scorn goodlife values and practices.

Further reflection on the don's proposal reveals ways the squatters and the Mexican Americans could have found common ground. Indeed, the don's plan begs the question: why would don Mariano propose a plan to the squatters that would enable their acquisition of land that was originally granted to his family by Spain and Mexico? One would imagine that he would be doing everything within his power to run the squatters off his lands, as don Santiago does in *Caballero*. However, the impulse to acquire and settle new territories resonates with him. His ancestors, who originally stole the Alamar rancho from American Indians, were taking part in settler colonialism, just as the squatters on the rancho were doing in territory they too considered "unsettled," or at least underdeveloped, as well. In his meeting with the squatters, don Mariano addresses them as equals, and he invokes their common ground. Later, he even says to Clarence, a squatter's son and his daughter's suitor, "I don't find it in my heart to blame those people [the squatters] for taking my land as much as I blame the legislators who turned them loose upon me" (Ruiz de Burton, *The Squatter* 161). Indeed, Clarence's many failed attempts to court don Mariano's daughter Mercedes result from the squatters' rejection of don Mariano's mutually beneficial plan.

However, the span of time that separates the settler colonial wave of Spanish and Mexican arrivals in California from the imperial/second settler colonial wave of American arrivals in California also represents differences in approach to settlement and cultural traditions. The Mexican American ranchers and the Anglo-American squatters do not readily recognize their commonalities as outsiders to California. The Spanish colonizers and Mexican settler colonials who arrived in California in the sixteenth, seventeenth, and eighteenth centuries sought to establish outposts, missions, and eventually a nation-state, in territories already inhabited by American Indians. Perhaps the opportunity for cooperation between Spanish settler colonials and American Indian communities existed in early encounters, but the historical record does not reflect this. The pattern of violent engagements, land takeovers, and imposed religious conversions in the Spanish settlement of the Americas is well-known. The Spanish invoked the same reasoning that the Americans intoned in later years about the Mexican Americans: the previous inhabitants did not adequately develop the resources of the land.

The Spanish colonization of the Americas initiated an ecological revolution from lower-impact American Indian horticulture and extensive burning to facilitate the growth of plants palatable to game animals to European-style ranching, including the introduction of species. As environmental historian William Preston, writing about environmental change in colonial California, observes about the American Indians in California:

> These peoples for the most part were not farmers so much as domesticators, and in a number of important respects all had domesticated portions of their habitat in order to induce greater subsistence. Instead of relying on the axe, rake, or plow, they skillfully learned to prune, till, coppice, transplant, and burn California's vegetation in order to encourage a greater abundance of plant and animal foods and materials. The land yielded its bounty to the ingenuity of native intervention and was substantially transformed in the process. (289)

That the colonial Spaniards and Mexicans were blind to the Indian domestication of California is clear. In the same way, Americans underestimated the Spanish and Mexican knowledge and influence over the land. One of the more significant changes was the way that U.S. imperial settlement of California initiated an ecological revolution from a land-grant system to a land-tenure system. This is a significant shift in scale from large landholdings managed by an elite class to smaller parcels of land managed by many more individuals and families of varying class status. This made an ecological impact, to be sure, although arguably this transition was bound to occur sooner or later as the population of North America steadily grew.

Of course there is some hypocrisy in Ruiz de Burton's critique: Spain and Mexico dominated American Indians in their own colonial takeovers of the territories they called home. Neither did they colonize a wilderness. This chapter reads these two novels in the context of settler colonialism, as well as in the contexts of imperialism, ethnic identity, and environmental history, to draw lessons from the unique perspective of upper-class nineteenth-century Mexican Americans who, on one hand, were oppressors and, on the other hand, anticipated the demise of their own privilege in the face of U.S. expansion. The upper-class Mexican American negotiation of race, especially claims to whiteness, in relationship to land possession and dispossession, clearly shows a perverse link between race and the ability to determine one's relationship to the natural environment under the aegis of coloniality, capital, and modernity. If ecocritics are looking for a way to access Mexican American contributions to environmental thought and for a way to understand the historical lens through which Mexican Americans view the natural environment, it would help if they understood Ruiz de Burton's two novels.

The Squatter and the Don makes a further argument in a feminist direction. Don Mariano's wife, doña Josefa, stands in the way of Clarence's romance with Mercedes because she does not want a squatter as a son-in-law. Unknown to her, Clarence buys the lands that his family inhabits from don Mariano, making him a landowner and no longer a squatter. Don Mariano and Clarence, however, decide to keep the women in the dark about the deal. Mercedes reluctantly rejects Clarence, obeying her mother's orders. Clarence, in the meantime, becomes quite wealthy due to wise stock market investments. If don Mariano and Clarence had informed doña Josefa of their dealings, and Mercedes had as a result been allowed to continue her relationship with Clarence, then the Alamar family would have taken a more positive turn alongside Clarence's charmed fate. This gives the novel a feminist twist, showing how leaving women uninformed and out of decision making creates misery rather than peace. The relationship between Clarence and Mercedes eventually blossoms, but only after much painful suffering and even deaths in the Alamar family. Once again, Ruiz de Burton parallels a young woman's life with the Mexican Americans' lives and their lands.

The real bit of wisdom to gather from Ruiz de Burton's depiction of this historical moment is the fact that Mexican Americans and their environmental knowledge were factored out of the land-management equation at a key moment of transition. They suffered a great deal of betrayal regarding this process at the initial moment of their becoming U.S. citizens and participating in U.S. democracy, that is, immediately following the U.S.-Mexican War. Indeed, don Mariano's dying words are not for love of family but for the heartbreak that his new nation caused him. Upon being asked by Mercedes if she can help relieve his pain, he whispers, "Too late. The sins of our legislators!" (Ruiz de Burton, *The Squatter* 304). That Ruiz de Burton wrote in the same era when the American Transcendentalists (including Ralph Waldo Emerson, Henry David Thoreau, and John Muir) were writing foundational environmental texts and when the United States was initiating a new era of environmental protection for the world is not an irony lost on Mexican Americans. Ruiz de Burton's work testifies to this fact, although it does so in the nineteenth century. Her observations remain relevant today. Goodlife writing's values and practices were manifest only in part at the time, with the concern for possession still of great importance to the landed aristocracy in Southern California. But the lessons learned from experiencing the oppression of possessiveness impact the ongoing cultivation of dignity and respect.

JOHN MUIR'S "DARK STRANGER" AND GOODLIFE VALUES

Preserving and communicating the Mexican American perspective on the environment is certainly not González's or Ruiz de Burton's primary goal in their novels, but their incisive social critiques cannot help but include this dynamic. Importantly, they also lay bare the groundwork that will continue to support the building of Mexican American views on the environment. Early Mexican American literary history offers a window into traditional agricultural and ranching methods, as well as important cultural practices that endeavored to keep human impact on local environments in check. However, their contributions have been overlooked.

Take, for example, the manipulation of the image of a young man that John Muir, nineteenth-century conservationist, nature writer, and Sierra Club founder, encounters on one of his hikes in the San Gabriel Mountains of California. The young man Muir encounters is possibly a Mexican American, perhaps a Mexican. He could have been born in Europe and emigrated to the Americas. Muir directly, if inconclusively, comments on his parentage: "his father was Irish and his mother Spanish" (*Steep Hills* 147). Muir seems to aptly represent the young man in his description of the encounter, though he never determines with certainty the young man's national origin or ethnic identity. I would not venture to definitively label the young man. However, a recent interpretation of the encounter reveals contemporary environmental studies' blind spot for Latina/o contributions to environmental knowledge. The story of the encounter is retold in an article by esteemed environmental historian Donald Worster. Worster paints a picture of friendly exchange across deep cultural divides between the two men, who find common cause in their "passion for nature" (par. 4). Yet in his narration of the encounter, Worster also deems it necessary to exaggerate the young man's foreignness, failing to recognize the possibility that the young man may descend from the generations of Spanish and Mexican presence in California. Moreover, in his emphasis on foreignness, Worster also seemingly overlooks the fact that Muir himself was an immigrant to the United States from Scotland. The encounter, taking place in 1877, is described by Worster:

> Along the way [Muir] came upon an *immigrant from Mexico* camped on the banks of Eaton Creek; predictably, Muir struck up a conversation with the *dark stranger* that lasted well into the night. In *halting English* his campfire host told about his dream of settling there amidst the oaks and chaparral, irrigating a vineyard and harvesting honey. *Since leaving his native land*, he had rambled a great deal—hunting, prospecting, and mining throughout the Southwest—but was now ready to make his home in this canyon paradise, to "make money and marry a Spanish woman." (par. 2, emphases added)

By 1877, California had been part of the United States for almost thirty years. This young man could very well have been born an American citizen, or, in any case, obtained U.S. citizenship by the mere act of staying in California after the 1848 Treaty of Guadalupe Hidalgo. There is reason to believe that he could not speak English very well; for many californios, Spanish was their primary language. However, the detail of the information that the young man conveys to Muir suggests that he has a pretty good command of English (or that, less probably, Muir spoke Spanish). It seems that Worster takes a lot of liberty in his description of this encounter. This license also, unfortunately, serves to erase the long historical presence of Mexicans and Mexican Americans in California and our diversity of experiences and appearances. I do not mean to argue that this young man was definitively of Mexican or Mexican American national or ethnic identity. I just want to point out how the possibility of associating this young man with a Mexican American history so easily dissolves to nothing in an environmental context.

Fortunately, Worster points to the specific passage upon which he bases his description. I quote it here, at length, to expose some of the significant inconsistencies, as well as to support my contention that Mexican American experiences and practices are oftentimes sacrificed to the image of conventional environmentalism:

> On the first day of my excursion I [Muir] went only as far as the mouth of Eaton Canyon, because the heat was oppressive, and a pair of new shoes were chafing my feet to such an extent that walking began to be painful. While looking for a

> camping ground among the boulder beds of the canyon, I came upon a strange, dark man of *doubtful parentage*. He kindly invited me to camp with him, and led me to his little hut. All my conjectures as to his nationality failed, and no wonder, since *his father was Irish and mother Spanish*, a mixture not often met even in California. He happened to be out of candles, so *we sat in the dark while he gave me a sketch of his life*, which was exceedingly picturesque. Then he showed me his plans for the future. He was going to settle among these canyon boulders, and make money, and marry a Spanish woman. People mine for irrigating water along the foothills as for gold. He is now driving a prospecting tunnel into a spur of the mountains back of his cabin. "My prospect is good," he said, "and if I strike a strong flow, I shall soon be worth five or ten thousand dollars. That flat out there," he continued, referring to a small, irregular patch of gravelly detritus that had been sorted out and deposited by Eaton Creek during some flood season, "is large enough for a nice orange grove, and, after watering my own trees, I can sell water down the valley; and then the hillside back of the cabin will do for vines, and I can keep bees, for the white sage and black sage up the mountains is full of honey. You see, I've got a good thing." All this prospective affluence in the sunken, boulder-choked flood-bed of Eaton Creek! Most home-seekers would as soon think of settling on the summit of San Antonio. (Muir, *Steep Hills* 147–48)

Here, Muir does indeed describe an encounter with a young Spanish-Irish man with whom he feels some commonality, just as Worster describes. However, Muir does not ever call the man an immigrant, nor does he intimate that the man could not speak English well. Quite the contrary, Muir reveals that most of their conversation occurred in the dark, as the man was out of candles. How could they have exchanged so much information with "halting English" and without seeing one another? Muir also indicates that he could not readily guess the man's nationality and eventually learns that his "father was Irish and mother Spanish." This combination was perhaps more common than Muir realized at the time. A young California woman, daughter of a Scottish father and an Indian mother, and foster child to a Spanish mother, figures as the title character in Helen Hunt Jackson's *Ramona* (1884), as just one example. Even so, we can excuse Muir's possible ignorance. He was, after all, born in Scotland, raised in Wisconsin, and only beginning his travels around California. Yet his unfamiliarity with the peoples of California does not prevent him from admiring the persistence of this young Spanish-Irish man, his campfire companion. The most significant observation in Muir's description of the encounter occurs at the end. Muir comments on the young man's extensive plans to settle an area that most people would find unattractive—comparing it to the nearby highest summit of six thousand feet, San Antonio! Muir does not go on to describe the young man's adaptability, but he clearly notes the depth of knowledge the man has acquired about the local flora and fauna, and admires his vision to make the most of an area that seems so uninviting.[3]

This chance encounter gathers significance over its repeated appearance in three different texts about the environment: Muir's *Steep Hills* in 1918, Worster's *A Passion for Nature: The Life of John Muir* in 2011, and now here. Muir retold his personal experience, which comprises a brief moment in a much longer and varied text, and in fact belies the image of him wandering the Sierras as a lone figure. Worster takes note of this encounter to demonstrate the ways that the "passion for nature" transcends the bounds of difference. I would like to bring attention to the way that this episode can help us see at least two things: the way that Muir takes note of this young man as ethnically unfamiliar (but not as Other) and still admires his environmental practices, and the way that, ninety-three years later, Worster interprets this man as an Other who originates beyond the national boundaries without taking into account the U.S. imperialism that brought about the recently changed political circumstances of the bioregion in which these two men met. Part of what makes Worster's interpretation possible is the common practice in the United States of erasing the long-standing presence and contributions of Mexican Americans, which is connected to legitimating the systematic dispossession of Mexican Americans of their lands.

Twenty-first-century ecocriticism would do well to engage Mexican American and Chicana/o literature as a source of insight, despite what some might see as a dubious background, to show how racism and environmental exploitation have been intertwined for more than a century. Mexican American authors rewrite the fundamental terms of environmental thought by emerging out of experiences of dispossession, poverty, and racism, fighting on the front lines of the ideology of Manifest Destiny and unlimited resources. Moreover, Mexican American writing reveals centuries of environmental care carried forward through channels of cultural hybridity and mestizaje. These three novels by Ruiz de Burton and González give us insight into how goodlife values and practices were partly shaped in response to the trauma of dispossession and discrimination. This trauma impacts the spirit and identity of the people as much as it determines their material fortune. Chapter 2 explores this aspect of goodlife writing by looking at the early twentieth century in the context of the "coloniality of being."

2

THE COLONIALITY OF BEING AND THE LAND

Identity in Early Twentieth-Century Goodlife Writing

I WOULD CHARACTERIZE the first literary historical period of goodlife writing as distinctly engaged with issues of material dispossession and discrimination, specifically in regard to land. The novels I discussed in the previous chapter, with their emphasis on domestic spaces, signal more intimate geographies, but public issues, such as economic and political power struggles over land possession and social position, remain their largest concerns. In this chapter, in which I loosely define the next era of goodlife writing, I explore more interior landscapes, especially coloniality's impact on human-nature relations in early twentieth-century goodlife writing.

In his essay "On the Coloniality of Being," Nelson Maldonado-Torres writes, "Colonization and racialization are not only political or social events or structures. They also have metaphysical and ontological significance" (260). These statements help elucidate a common critique of early twentieth-century Mexican American writing. Critics have accused these writers of lacking political edge and social critique, and some go as far as accusing them of selling out or romanticizing Mexican culture. These works are therefore sometimes taken to be irrelevant to the concerns of Chicana/o studies. In an essay that offers an intriguing analysis of Mexican American literary responses to the American Dream, Raymond Paredes comments on the early twentieth-century period of Mexican American literature:

> Writers in English, Cristina Mena as an example, gave their readers stoical peones and fiery señoritas who lived in tightly knit communities on both sides of the border and who seldom bothered to think about such indelicate issues as material advancement. These romantic literary Mexicans looked to the past instead of the future and desired nothing so much as to be left alone. In short, they were the types of Mexicans that Anglos liked best: primitive, colorful and, most of all, unthreatening; and insofar as they understood the myth at all, they regarded the American Dream as simply another feature of a culture they steadfastly rejected. ("Mexican American Authors" 74)

This critique sounds accurate in some ways. In 1936 Nina Otero Warren published a book titled *Old Spain in Our Southwest*, where she documents traditions of Northern New Mexico Hispano culture such as "Asking for the Bride" and "Saints Days and Feasts." Eva Wilbur Cruce and Fabiola Cabeza de Baca each wrote a memoir of her childhood on her family's ranch. Jovita González gathered the folktales of her South Texas ancestors. Sabine Ulibarrí published a short story about a man who forgets his own name and is haunted by his father's voice. Explicit antagonism between Mexican and Anglo culture escapes mention in these works. What does merit concern is the state of the soul told through memoir, personal essay, folktale, and short story. Given the metaphysical and ontological significance of colonization and racialization, these works gather a new significance and relevance to today's ongoing coloniality of being.

A curious shift occurs in the transition from late nineteenth-century to early twentieth-century Mexican American writings. While late nineteenth-century writers, such as María Amparo Ruiz de Burton, center their writings on the ways Mexican Americans can fit into the U.S. social landscape and power structures, and writers from the mid-twentieth century until today largely center their works on civil rights and social injustice, the early twentieth-century writers retreat into interior terrains. The first Chicana/o literary scholarship labeled the early twentieth-century period of Mexican American writing as devoid of politics, while more recent studies demonstrate radical aspects of this literature, some even arguing that the progressive politics in these works rival those published during the civil rights era.[1] Tey Diana Rebolledo declares that this era combines "a history of resistance and accommodation to the social, economic, and cultural hegemony of the white Anglo-Saxon people who came to dominate land and society in the Southwest" (*Women Singing* xviii). In this chapter, I argue for yet a third approach, distancing myself altogether from the need

ation: the land. By giving the land a spirit with which they connect, riters reject the colonial objectification of the Amerindian, the mes- d the natural environment. This is a subtle restructuring that addresses onized subject's inner process of decolonization—what Mignolo calls ng—more than an external deconstruction of power, but it is no less ul and vital a process.

negotiation of a racial hierarchy was not a new experience for Mexican cans. The situation was not very different from the one they found in al and postcolonial Mexico. Mexico, including the territories ceded to nited States in 1848, had a racial hierarchy that resulted from Spanish ization. The difference was that Mexican Americans in the United States themselves decidedly on the downside of a racial binary: those who t have been white or mestizo in the Mexican caste system had no choice be colored in the United States. As descendents of the losing side of the war the United States has fought in territory contiguous with its own, ican Americans possessed little cultural capital. Their claims to large tracts nd, as discussed in the previous chapter, also put them in the sights of -hungry settlers, contemptuous of Mexican American territorial claims, ing from the eastern United States. These writers capture this new nego- on of identity by writing not just of themselves but of the land as well.

During this period in Mexican American history, access to any significant ition of influence, and certainly any claim to land titles, entailed a strategic m to whiteness and a corresponding denial of indigeneity. Looking at writ- s from this period through a decolonial goodlife lens helps to better under- nd how the negotiations of identity have largely been misunderstood as a rayal of Mexican American heritage and cowardice in the face of discrim- tion. Goodlife writing from the early twentieth century reveals another dimen- n of the ongoing process of delinking from coloniality and in so doing offers aders much-needed insights into the trauma inherent in the decolonization/ linking processes.

The impact of coloniality made it difficult indeed to forge a Mexican Ameri- n identity in the early twentieth century. I implement Paula Moya's definition f "identity" here as "the nonessential and evolving product that emerges from he dialectic between how a subject of consciousness identifies herself and how thers identify her" ("Who We Are" 80). This definition of identity gives both he self and the social context a role to play in the ongoing process of defining dentity. Moya's understanding of identity does not privilege individuality over community, nor does it emphasize the group over the self. Importantly, this is not only a dialectic, it is an ongoing process. Early twentieth-century Mexican Americans found themselves pulled in different directions by their own Mexican American context that allowed them space to define themselves against the still relatively new U.S. setting, which imposed a perception of Mexican Americans as undesirable, mongrel, and incompetent. At the same time, the self depicted in these works sought to maintain at least one element of continuity—relationship with the land—that could also help individuals negotiate external pressures of perception. It is here that this goodlife writing analysis of these works makes its original contribution.

Some writers, such as Adelina Otero Warren of New Mexico and Eva Wilbur-Cruce of Arizona, made strategic claims of whiteness in order to gain access to social position and land, but they also verged on reproducing the very oppression that they sought to escape. Their writings reveal an awareness of and a form of regret about their strategic adjustments to modernity/coloniality. Other writers, such as Sabine Ulibarrí of New Mexico, focus more on the feelings of duplicity and betrayal produced by the pressure to claim a limited identity for the sake of power and influence. His protagonist is literally haunted by his ambivalence, constantly hearing his dead father's voice. It seems that only a few writers, such as Jovita González of Texas and Fabiola Cabeza de Baca of New Mexico, were versatile enough to claim their mestizo identities in recognition of both their American Indian and their Spanish heritage. Although by present-day standards, a claim of mestizaje may not seem to completely disengage from modernity/coloniality, their communion with the land proves their decoloniality. All of these writers communicate their ongoing adaptations of Mexican American identity and the ways that coloniality endeavors to stand between this community and the lands they had known for generations. This period of tentative delinking proves fundamental to understanding the persistent strategies Mexican Americans employed to ensure their connection with the land. In affirming their relation with the land, they deconstructed the idea of soulless subjects of coloniality, be those subjects "nature" or "native."

Instead of delving deeply into any one or two works from this period, I offer a number of exemplary works in order to demonstrate a general pattern and a range of approaches that typify the era. I also define this period of the "early twentieth century" broadly: the beginning of the twentieth century to the Chicana/o Civil Rights Movement. I discuss two ranch memoirs together, Eva Wilbur-Cruce's *A Beautiful, Cruel Country* and Fabiola Cabeza de Baca's

We Fed Them Cactus. From there I will move to the introduction Adelina Otero Warren wrote for her book, *Old Spain in Our Southwest*. Another folklorist, Jovita González, offers a treasure trove of stories from South Texas that appeared in the Publications of the Texas Folklore Society. Finally, I will offer a reading of a story from Sabine Ulibarrí's collection *Tierra Amarilla*, titled "The Man Without a Name." All of these writers make visible the struggle of delinking from coloniality during the time period that preceded the era of militant Chicana/o civil rights.

The connection between racial identity and the natural environment has heretofore been little explored. In his discussion of whiteness in nineteenth- and early twentieth-century Mexican American literature, B. V. Olguín argues that "the performance of Mexican American and Chicana/o mestizaje has rendered their indigeneity *sous rature* (or under erasure)" (32). Current studies in critical indigeneities insist that Chicana/o studies must examine its "nostalgic indigenismo" if it wants to remain relevant. Olguín goes as far as calling nostalgic *indigenismo* a specter. Much can be learned from this challenge of rooting out the ghost that haunts Chicana/o studies. At this critical juncture, we can learn more from the way Mexican Americans negotiated this difficult territory in the early twentieth century. These five writers in particular present part of the problem of nostalgic indigenismo at the same time that they affirm the idea that both they and the land possess souls and qualify as entities, thereby deconstructing the insidious coloniality undermining their being.

MEXICAN AMERICAN RANCH MEMOIRS: LONELINESS WITHOUT DESPAIR

Both Fabiola Cabeza de Baca and Eva Wilbur-Cruce invoke their experiences in the rough and rural Southwest landscape during the early twentieth century. Cabeza de Baca, in *We Fed Them Cactus*, narrates her childhood on an eastern New Mexico cattle ranch, where her father let her stray from conventional feminine behavior by allowing her to ride horses and explore the ranch that spread onto the vast expanses of the llano. Wilbur-Cruce, in *A Beautiful, Cruel Country*, tells of a difficult childhood in southern Arizona, where her father expected her to contribute at least as much work as a ranch hand. She rode and wrangled with the men when she was only five years old. The narratives that emerge from these unique experiences reveal women—their voices, characters, and careers—shaped by the extremes of the U.S. Southwest. Over the course of their lives, these women take the arid climate and threatening animals in stride as they labor, in awe of the beauty that surrounds them. Their time and place, they argue, gave them the rare opportunity to upend conventional gender roles and question the capitalist accumulation of wealth. Rather than feel alienated from the harsh environment, they felt in communion with the land, water, plants, animals, and climate that sustained their everyday lives. These memoirs honor the spirit of the places that shaped their childhoods.

Both memoirists set out to write about the place where they grew up more than about the sum of their childhood and adult experiences. This makes these memoirs more about one's relationship to place—to a particular bioregion—than they are memoirs of individual women. Cabeza de Baca opens her narrative with a lyrical description of the Llano Estacado, the Staked Plains of eastern New Mexico and western Texas: "The Llano is a great plateau. Its sixty thousand square miles tip almost imperceptibly from fifty-five hundred feet above sea level in northwest New Mexico, to two thousand feet in northwest Texas . . . [T]he vegetation includes juniper, piñon, yucca, mesquite, sagebrush, gramma and buffalo grasses, as well as lemita, prickly pear, and pitahaya" (1–2). This is where her family held land for many generations, dating back to a Spanish-era land grant, but the reader does not discover this information until after she reads an in-depth description of the place and its effect on the soul: "It is a lonely land because of its immensity, but it lacks nothing for those who enjoy Nature in her full grandeur. The colors of the skies, of the hills, the rocks, the birds and the flowers, are soothing to the most troubled heart. It is loneliness without despair. The whole world seems to be there, full of promise and gladness" (3). Here she first indicates the way one can learn from the Llano—to inhabit a sense of loneliness without despairing, but only if one cultivates a capacity to "enjoy Nature in her full grandeur."

Wilbur-Cruce also makes a point of opening her narrative with a description of the landscape's impact on her spirit: "[M]y isolation has always been only physical, not spiritual. I have never felt depressed or lonely when alone with the land. Something always happens that dazzles me and overwhelms me with amazement and wonder" (vii).

Wilbur-Cruce's narrative takes place in the southern part of Arizona, outside Tucson and near the U.S.-Mexico border at Arivaca Creek. Her family's sixteen-square-mile ranch—established by her grandfather, Wilbur, who homesteaded it in the mid-nineteenth century—reached all the way to the border.

By the time she published her memoir, Wilbur-Cruce was in her eighties, but the portion of her life that she chose to narrate were two years full of formative experiences: from age three to age five. One would not expect such a young child to have experienced so many memorable adventures, but Wilbur-Cruce is certainly an exception. The way she frames her narrative—bookended at the start with a climb up a mountain and at the finish with the government's relocation of her Tohono O'odham friends—communicates the significance she gives these events. They are the beginning and end of her most vital experiences.

That both memoirists make a point of sharing their experiences of "loneliness without despair" or depression and the rewards of solitude might prompt one to make comparisons of these works with John Muir, Henry David Thoreau, or even Edward Abbey. The communion with the natural environment in the works by these two Mexican American ranch memoirists differs significantly from the works of these iconic American nature writers. Cabeza de Baca and Wilbur-Cruce reveal a community relation with their places, always including the local peoples as integral to the way they experience their environments. Their knowledge and experience of the Mexican American Southwest defied conventional gender roles and upended economic values. They achieved all of this through solitude in the landscape and solidarity with the land, weaving details about their individual lives with the contours of the landscape, thereby showing their readers the ways they developed personal traits from their experiences in the arid Southwest.

Fabiola Cabeza de Baca was only four years old when her mother died, and she developed an especially close relationship with her father. He let her roam the ranch and take part in physical activities, such as riding and roping, that girls did not typically perform in that time and place. Her feeling of liberation from typical gender roles led her to question conventions in off-ranch settings as well: "Food was served after the Mass, and the old custom of feeding the men first still prevailed. I always sat by Papa and the women and girls eyed me with curiosity" (134). Why should she wait to eat with the women, after the men? Her youth of crossing gender-role boundaries inspired Cabeza de Baca to lead an independent life. Wilbur-Cruce also rejected traditional gender performance. In her case, her mother helped her along by sewing custom "mini-Levis" for her in the days when "Levi Strauss did not manufacture pants for women, much less for children" (175). "Cowgirls wore divided skirts, and not without criticism. Such ugly comments as 'I'd like to stone her to death' were not uncommon" (175). Young Eva endured community censure for the sake of her love for the family ranch, and she eventually took over the ranch in 1928—certainly a time in which few twenty-eight-year-old Mexican American women ran ranch operations.

Both memoirs also challenge conventional economic values. Rather than seeing liquid cash or the accumulation of material goods as wealth, they both sought money and objects only on as-needed basis. Cabeza de Baca says, "All the rooms were spacious and our home had a feeling of hospitality. We had only the most necessary pieces of furniture" (10) and "We had never been poor, because those who live from the land are never really poor, but at times Papá's cash on hand must have been pretty low" (11). Here Cabeza de Baca deconstructs the category of poverty itself, declaring that intimacy with the land excludes the possibility of poverty altogether. The only poverty possible pertains to the health of the land. Given the arid settings of both memoirs, it is not surprising to discern that real wealth was to be found in rainfall. Cabeza de Baca firmly declares, "Money in our lives was not important. We never counted our money; we counted the weeks and months between rains. I could always tell anyone exactly to the day and hour since the last rain . . . Rain for us made history" (11–12). Wilbur-Cruce describes the difficulty of life in southern Arizona and the difficulty with which "abundance" was acquired: "[W]e had to blast it out and grind it and wash it, and . . . finally, the 'gold nuggets' . . . appeared, in the guise of hope, beauty, joy, love, and brotherhood, after which all things became bearable" (xii). Both writers reject the capitalist accumulation of material wealth in favor of sustenance and care for the land in a reciprocal relationship.

Both writers dedicated their lives to sustaining the land-based culture in which they were raised. Cabeza de Baca attended New Mexico State University and earned a bachelor's degree in home economics in 1927. She put her degree to practical use in a thirty-year career as an agricultural extension agent. Her work is emblematic of epistemic disobedience, as she used her government position to create institutional space for and preservation of New Mexican culture and knowledge. In addition to her memoir, *We Fed Them Cactus*, she also wrote two cookbooks that not only list recipes but also detail cultural and environmental practices that accompany the foodways she documents. But her signal contribution, overlooked until now, involves her communion with the land in a way that rejects the structure of conventional nature writing.

Wilbur-Cruce became a ranch owner—a pursuit almost exclusively occupied by men—and her ranch was famous for preserving the "rock horse" breed that she writes about in her memoir. She took up the pen only late in life, publishing *A Beautiful, Cruel Country* in 1987 and also participating in the oral histories recorded in Patricia Preciado Martin's *Songs My Mother Sang to Me*.

ADELINA OTERO WARREN'S NEW MEXICO AND OLD SPAIN

Adelina "Nina" Otero Warren was from northern New Mexico, a region that engenders a rich history of writing about the natural environment. Her writings at this unique place and time communicate her ambivalence about her identity via the environment. Given that Mexican Americans developed the "Spanish American" identity in part as a means to maintain control over landholdings, it cannot be considered a coincidence that Otero Warren uses the natural environment to express her reservations about the Spanish American identity. I first offer a bit of background about Otero Warren and a brief history of the development of the "Spanish American" identity as a prelude for demonstrating how her introductory essay in *Old Spain in Our Southwest* expresses her reservations about that identity and her frustration with the prospect of her culture losing its intimacy with the natural environment in New Mexico. Otero Warren shows that this transition in distancing Mexican American identity from the natural environment took its psychic toll, and her ambivalence about it reveals her own struggle with the coloniality of being.

Nina Otero Warren—prominent citizen, leading educator, and political leader in Santa Fe during the early twentieth century—published only one book in her lifetime, *Old Spain in Our Southwest*, in 1936. Her interest in the Spanish impact on New Mexican culture no doubt emerged from her heritage. She was born in 1881 in Los Lunas, New Mexico, as part of an extended family that claimed, on her mother's Luna side, ancestry from the Spanish who colonized New Mexico in the sixteenth century and, on her father's Otero side, from the eighteenth-century Spanish settlement of New Mexico. At the time of her birth, New Mexico had been a U.S. territory for thirty-three years, but it still was not a state. She was born and raised an American, but in the context of a traditional Mexican family that aspired to be included in the upper echelons of the changing power structure of the region.

She is most noted for successfully leading the New Mexico suffragette movement from 1915 to 1920 and for her nomination to run for the U.S. House of Representatives in 1922.[3] And even though she was the first woman in New Mexico to compete for this office, she lost by only 9 percent of the overall vote. She is also known for serving in the position of superintendent of schools in Santa Fe County from 1917 to 1929 and for being the first woman to serve as inspector of Indian services for the Department of the Interior from 1922 to 1924. Because she was a prominent citizen in her day, with influence over local policy and practices, her writings, although limited to one book and a few articles, can reveal a lot about the social group and class to which she belonged, and about the strategies she used to achieve her goals of preserving and protecting the Spanish American culture of her time, including the Spanish American relationship to the natural environment.

Otero Warren identified as Spanish American, as did many Mexican Americans of the late nineteenth and early twentieth century in New Mexico. This is not a simple matter of interchangeable labels. Studies by John Nieto-Phillips and Laura Gómez indicate that Mexican Americans in the late nineteenth and early twentieth century identified as "Spanish Americans" to indicate their whiteness, even as the majority of Anglo-Americans considered Mexican Americans an inferior and mongrel race. In the case of New Mexico, the Mexicans-turned-Americans attempted to blend something of their past elite status, "Spanish," with their newfound citizenship, "American." Still, it is an irony that Mexican Americans had to develop a strategy to claim whiteness at all once they became Americans. Under the terms of the Treaty of Guadalupe Hidalgo, all Mexicans in territories acquired by the United States following the U.S.-Mexican War were eligible for U.S. citizenship, and only "whites" were allowed U.S. citizenship at the time. However, popular sentiment did not accord with the legal terms. According to Gómez, "a broad consensus existed among Euro-Americans that Mexicans were nonwhite precisely because they were racially mixed. For many Americans, it was the fact of Mexicans' 'mongrel' status that most strongly signaled their racial inferiority" (83). Such negative public sentiment impacted New Mexico's efforts to become a state. A Congressional report in opposition to New Mexico's statehood in 1876 notes that the majority population of the territory consists of "a mixture of Spanish or Mexican and Indian in different degrees" (Nieto Phillips 69). The report further claims that "the inhabitants of this isolated region, with few exceptions, continued to sink, till now for nigh two hundred years, into a condition of ignorance, superstition, and sloth that is unequaled by their Aztec neighbors, the Pueblo Indians" (69). In the context of such perceptions about Mexican Americans of the region, Nieto-Phillips explains the strategy of a new label:

> As *Spanish* Americans, Nuevomexicanos possessed a readily understood racial identity, a documented historical lineage, and a claim to the land that dated to the very "conquest" of New Mexico, beginning in the sixteenth century. In a rhetorical sense, then, this term restored their presumed "Spanish blood" and, by implication, rendered them racially "fit" for—and deserving of—self-government and "American" civic rights and responsibilities. (82)

I would add that Otero Warren's writings suggest that part of the "claim to the land" that the Spanish American identity offered was several centuries of knowledge, experience, and stewardship of the local environment. Yet, even when implementing such strategies that recalled a long historical presence on the land, it took the United States sixty-four years, from 1848 until 1912, to make New Mexico a state in its union.

Yet, as Gómez shows in her far-reaching study, Mexican Americans lost some valuable ground in their unilateral claim to Spanish heritage and desire for power alongside Anglo-Americans. In this regard, they played into an Anglo-American ploy to "divide and conquer." By claiming whiteness, the Spanish Americans were also claiming superiority over local indigenous groups, including the Pueblo Indians, with whom they had worked in the past. These superiority claims, of course, kept them from the possibility of allying with Pueblo and other indigenous groups—as a majority population—in New Mexico *against* Anglo-American quests for power and land. Indeed, their claims to whiteness "legitimiz[ed] the American presence as 'protector' of Indians, and entrench[ed] the American legal system as a neutral, fair forum for dispute resolution and punishment" (Gómez 114). In addition to implicitly approving the American judicial system and mistreatment of Native Americans by, in essence, aspiring to whiteness, the Spanish Americans also missed an important opportunity to resist the reinforcement of a racist hierarchy at this time. As Gómez sharply observes:

> The power of racism is ideological, achieving its apex when racially subordinated groups themselves help to reproduce racism . . . [T]he ability of Mexican Americans to at times succeed in claiming whiteness led them into a perverse trap. To

> solidify their classification as white, they had to act like whites, especially with respect to non-white groups. Mexican American elites, in particular, acted in ways that shored up their whiteness, at the expense of every non-white group below them in the racial hierarchy. Intentionally or not, they became agents in the reproduction of racial subordination and contributed to the consolidation of a new version of white supremacy in the Southwest. (113, 115)

By the time Nina Otero Warren began her political life in 1915, at the age of thirty-four, she and her elite family were well established as Spanish Americans with some influence in the region and recognition and respect from Anglo-Americans. Her second cousin, Miguel Antonio Otero II, was territorial governor from 1897 to 1906, and Otero Warren spent her youth socializing in these powerful circles in Santa Fe. There is no denying that her families on both sides—the Lunas and the Oteros—ended up making a play for whiteness during the transition from the Mexican to the U.S. era in New Mexico. She finds herself in the second generation of Spanish American identity, and her writings in particular, along with her political and professional activities, are an attempt to make sense of what culture this identity professes to preserve and what utility it can offer. I will offer evidence that Otero Warren signals ambivalence about her Spanish American identity in her 1936 book. Perhaps Otero Warren felt and regretted the disadvantage for others that her own privilege created, and she expressed this not through a loss of connection with other peoples but through a loss of connection to her local places.

The ambivalence she shows in her introductory essay presents a paradox because the book was meant to fill the gap in educational materials available to Spanish American youth in the schools. Her experience as superintendent of schools in Santa Fe County made it clear to her that Spanish American youth could benefit from learning about their own culture in an institutional setting—until then, they only learned about the Anglo-American culture and traditions in school. *Old Spain in Our Southwest* addressed this concern. The chapters capture different aspects of hacienda life and traditions, with titles such as "An Old Spanish Hacienda," "Asking for the Bride," "The Harvest," and "Spanish Place Names." It was well received at the time, as indicated in short reviews in various publications, though the longer reviews considered it a predictable and somewhat romanticized celebration of Spanish American culture. In 1938, Otero Warren's contemporary, A. L. Campa (a well-known folklorist and one who disputed the Spanish American label), wrote, "The book is far too subjective and it emphasizes the spiritual life of New Mexican culture" (209). In 1978, Chicana/o literature critic Raymund Paredes found the book "profoundly disturbing" and a "literature created out of fear and intimidation, a defensive response to racial prejudice" ("Evolution" 87).

Given the context of recent colonization and the evolution of Spanish American identity as a play for political power in the Anglo-American era, both Campa's and Paredes's observations are apt. However, Otero Warren's writing offers more. In 1989, Chicana literary critic Tey Diana Rebolledo notes that the book was an "early attempt to preserve in literary images a vanishing way of life" ("Las Escritoras" 202). She also notes that Otero Warren "communicates to us her growing sense of disjunction from the landscape, illustrating her feeling of alienation and isolation in a time of transition" (203). Considering these past evaluations of Otero Warren's writings, especially Rebolledo's suggestion that Otero Warren felt alienated from her landscape, I found myself drawn to the opening essay and wondering how the coloniality of being could produce a new or different evaluation of a seventy-nine-year-old text. How does Otero Warren make visible her ambivalence about the decline of her communion with the land?

The opening chapter of *Old Spain in Our Southwest*, "The Wind in the Mountains," offers a rich opportunity to look at how she expresses her connection to the land and her sense of loss. This is also the part of the book that drew Rebolledo's attention in evaluating Otero Warren's writing.[4] This opening chapter remains unique in the overall work, taking a less pragmatic and descriptive tone than the rest of the chapters. Otero Warren seems to want to intrigue her readers at the outset, offering an enigmatic opening in which to contextualize the more straightforward descriptions of Spanish American life. I would even venture to say that the opening chapter is addressed to a more sophisticated audience (the teachers, reviewers, critics) while the remainder of the text is addressed to the adolescent students who have yet to learn of their own culture in school. The fact that I have found several revised drafts of this chapter—many more than for the other chapters—of the book in her manuscript collection at the New Mexico State Archives shows that Otero Warren took extra pains crafting this piece, apparently putting more work into it than into the others. It is a mere three pages of the overall book, but a close examination of its contents and revisions offers a great deal. The chapter opens with her first-person description of her natural surroundings and an experience of a storm one night in her "small adobe house in the midst of cedars on the top of a hill" (3). She is alone in the house, but she observes a shepherd (doubtless a mestizo) and "an Indian" who weather the elements outside through the night while she remains indoors. Here is the closing paragraph, to give you a feel for the piece:

> As the shepherd was extinguishing the camp fire, there appeared on the top of the hill a form with arms stretched to heaven as though offering himself to the sun. The shepherd from his camp and I from my window watched this half-clad figure that seemed to have come from the earth to greet the light. A chant, a hymn—the Indian was offering his prayer to the rising sun. The shepherd, accustomed to his Indian neighbors, went his way slowly, guiding his sheep out of the canyon. The Indian finished his offering of prayer. I, alone, seemed not in complete tune with the instruments of God. I felt a sense of loss that they were closer to nature than I, more understanding of the storm. I had shuddered at the wind as it came through the cracks of my little house; now I had to cover my eyes from the bright rays of the sun, while my neighbors, fearing nothing, welcomed with joy "another day." (5)

There is a wistful tone here that never reappears in the sketches about traditions and customs that follow. She even betrays envy for the freedom and comfort (in nature) that her neighbors the shepherd and the Indian enjoy. Rebolledo notes that Otero Warren "remains on the edge, longing for the sense of integration denied her, in part because she is a woman, but also because she represents a culture and class undergoing a profound transition" ("Las Escritoras" 203). A woman of her class could hardly admit to envying the lower classes, for Spanish Americans sought distinction from the American Indian and the mestizo, as I have noted above. However, Otero Warren still expresses wistfulness about the passing on of intimacy with the natural environment. After all, it is around this time that writers and artists travel to New Mexico in search of authentic experiences in nature and with "natives" of the Mexican as well as of the American Indian variety. However, she does more than simply romanticize. She mourns the passing of a time that allowed for the possibility (at least) to work together with American Indians and to identify more closely with mestizos. In this mourning, she protests, if in a veiled way, the dominance of Anglo-American culture, which precipitated an irreversible separation of the Spanish American from the American Indian and the mestizo—and also from the natural environment, for, as I point out in other chapters of this book, the story of Mexican Americans is a story of strong connections with the natural environment through goodlife work and also work in both a natural and a socioeconomic environment severely degraded by modernity/coloniality. In other words, with the brief introduction to her book that works to preserve Spanish American traditions, she laments the fashioning of the Spanish American identity precisely because it forced her to leave behind the possibility of intimacy with mestizos, Indians, and the natural environment. Even so, she is practical and continues the project of the rest of the book. But it cannot remain unrecognized that she does so with an initial expression of ambivalence, even critique.

Analysis of her drafts of this opening chapter offers evidence of her process of expressing communion with the land through her cultural identity. Her original draft incorporated Spanish words throughout, including the language of the title; the published version retains none of the Spanish from the original. The title from the draft was "El Aire en las Montañas," but the published version is "The Wind in the Mountains." A major debate in education at the time was, of course, about bilingual instruction. Otero Warren consistently argued in favor of bilingual education and, in the course of this essay, offers a sneaky barb at those intimidated by an unfamiliar language: "The night was alive with sounds of creatures less fearful than humans, speaking a language little understood by us yet felt by every sense" (*Old Spain* 4). This critique of those unable to "feel" with "every sense" the unfamiliar language of animals made its way into the final draft, but remains enigmatic unless one contextualizes it within her larger critique of the increasing domination of Anglo-American culture and English in New Mexico. She admires these creatures for being less fearful than humans and allows herself to "feel" with "every sense" their unfamiliar language—sounding a challenge to others facing an unfamiliar language. However, another less enigmatic warning to outsiders did not make the cut. Early in the essay, when she describes the coming of the storm, she also warns: "This great Southwest is beautiful beyond words, but it needs to be watched lest it bruise those not acquainted with its moods" ("The Wind in the Mountains"). Here she sounds a warning to outsiders about the unpredictability and force of local weather patterns, but she might be suggesting the power and will of its people as well.

Indeed, the trope of the storm itself represents the American colonization of the Southwest. Several points in the essay indicate this. First, she describes distinct feelings before and after the storm. Before the storm she had a "feeling of vastness, of solitude, but never of loneliness" (*Old Spain* 4). However, after the storm she "felt a loneliness not apparent in those more understanding of the aftermath of the storm" (5). The onset of a feeling of loneliness comes only after the storm of colonization, when she takes up a Spanish American identity and its associated schism with Indians, mestizos, and the natural environment. Even so, she is determined to depict a persevering culture—which she likens to a candle in the storm. Before the storm, she describes "a melancholy candle . . . flicker[ing] as if gasping for breath before expiring" (4), and after the storm we find that the candle's flame has survived: "An unmuffled candle alone illuminated the small room. It kept vigilance through the stormy night" (4). She realizes that something of her original culture, which valued at least some of the mestizo and Indian cultures, can still survive this colonization that has distanced her from them and from the natural environment. Perhaps she is signaling that it is these aspects that she attempts to preserve in her book for the coming generations.

One cannot deny the oppression that the Spanish conquistadores and their descendants enacted on the indigenous peoples of the Americas in the process of colonization, nor can one deny the strict caste system that emerged from the Spanish conquest. This also seems to be hinted at in Otero Warren's opening chapter, in its published as well as in its draft form. In describing the effects of the storm, she wrote two versions. I present both, the published version followed by the draft version:

> The wind hissed like a rattler, and as it struck the branches of the trees, it made a weird sound like a musical instrument out of tune. Trees were bowing as if in obeisance to their Master. (4)

> The wind hissed like a rattler before it strikes. We arose to see the trees bowing, as if in obeisance to their Master. Down crashed the dead branches, the wind striking the tips of those more obstinate with the weird sound of a musical instrument out of tune, like the crash of drums and symbols [*sic*] emphasizing a strain of music while an idol fell to the ground. ("The Wind in the Mountains")

In both these versions, she clearly states the way the storm controls the situation and changes the landscape, with the trees bowing to their "Master." With the use of the word "Master," she indicates a shift in power, both recognizing the new masters and acknowledging that another class of "Master" prevailed before, and that the trees resist the shift in power with their "out of tune" music. She notes a resistance to the American colonization of Mexico, but she also recognizes the success of the United States. However, the unpublished version makes this point more emphatically by calling some of the tree branches "more obstinate" and observing that their fall sounded "like the crash of drums and symbols [*sic*] emphasizing a strain of music while an idol fell to the ground" ("The Wind in the Mountains"). Why does the falling of the obstinate tree branches crash like drums and "symbols"? Why do the tree branches fall to a tune that might play when an "idol" falls to the ground? In this draft version, she emphasizes the significance of the passing of Mexican power to Americans (crash and drum). I would also argue that she suggests that the Mexican era's fall was like the fall of an idol—and she did not mean an "idyll," because the Mexican era was far from idyllic—and in so suggesting, she is critiquing the weaknesses of that era, including its entrenched peonage/slavery system.

In *Old Spain in Our Southwest*, Otero Warren records the passage of the Mexican era, and does her best to preserve a memory of it for New Mexican students. But she also critiques the old way, calling it an "idol" in the draft version of her opening essay. If we are to take this opening essay as a context for the rest of the book, then we must read her ensuing comments in the light of her new "loneliness" and alienation from the natural environment. When, in the chapter describing life on the "Old Spanish Hacienda" she says, "Never did anyone but peones expose themselves to the sun" (9), we should remember that she laments this distance from the elements. An environmental reading of this work offers a more complex understanding of this Spanish American woman who advocated on behalf of Spanish Americans, American Indians, and mestizos in her lifetime, and who did not forget the lessons of the past. Her writings lamented the compromises of the Spanish American identity made for the sake of power under the American system, but she was unwilling to let students remain without some knowledge of past practices. In this sense, she emerges as a voice of ambivalence regarding her Spanish American identity and what it cost her in communion with the land.

JOVITA GONZÁLEZ'S MESTIZAJE AND THE SOUTH TEXAS ECOLOGICAL REVOLUTION

While Nina Otero Warren documented life in northern New Mexico, Jovita González recorded the folktales of South Texas. Rather than communicate ambivalence, González registers a cryptic resistance found in traditional stories.

The folktale genre keeps the tone light, but the tales that González chooses to record ultimately communicate a critical response to modernity's interference with the Mexican American relationship with the land. González tells the story of modernity's intervention by documenting the early twentieth-century ecological revolution of South Texas. "Ecological revolutions," as defined by environmental historian Carolyn Merchant, "are major transformations in human relations with nonhuman nature. They arise from changes, tensions, and contradictions that develop between society's mode of production and its ecology, and between its modes of production and reproduction. These dynamics in turn support the acceptance of new forms of consciousness, ideas, images, and worldviews" (2). In some of her stories and in her novels, González responds to the ecological revolution in her homeland by seeking to preserve a way of life and the environmental knowledge that accompanied it, which she saw die out. Especially significant in this regard are her first published stories—folktales collected in the 1927 edition of the Publications of the Texas Folklore Society, *Texas and Southwestern Lore*, edited by J. Frank Dobie. These stories, each with its explanation of a plant or animal trait—like the dove's coo, the Texas sage's ashy color, the mockingbird's white wing feathers, the thornless thistle, and the Guadalupana vine's medicinal value—challenge readers to see the critique hidden in their message of traditional environmental knowledge. This represents a hitherto unrecognized source of environmental writing, emerging from the diligence of Chicana/o scholars to recover Mexican American writing from this historical period. Although at first glance these tales may seem simple and lighthearted, they prove to be intricately disguised and serious critiques of the large-scale cultural and agricultural changes that González and her community witnessed in the area.[5]

The changes that González documents were brought about by three related and powerful forces: the growing American influence after the U.S.-Mexican War, the 1904 arrival of the railroad, and the resulting shift in the region from ranching to farming. The writings of this South Texas native daughter document the transformation of the greater Rio Grande Valley in the late nineteenth century through the early twentieth century, paying close attention to the way these changes affected the human relationship with the natural environment. The shift that she witnessed meant more than a change in land use. It also meant a power shift that moved her community from a culture of landowning Mexican American ranchers who afterward found themselves to have become landless laborers.

By the time González was an adult, the halcyon days of grand Mexican American ranchos in South Texas were mostly a memory. Her own ancestors had once been ranch *patrones*, numbering among the border aristocracy. And although she did not inherit a grand ranch, she took seriously the cultural inheritance that was available to her: stories of her ancestors and culture. González also focused her attention on the humblest population of Mexican Americans, the peones, honoring them by rooting her writings in the folktales they shared with her. She was always careful to include stories that narrated the Mexican American relationship with the natural environment. Interestingly, González introduces her stories by giving credit to "the quaintness and simplicity of the Indian myth" ("Folklore" 7). This may sound condescending to contemporary ears, but it is important to note the frankness with which González declares, many years in advance of Anzaldúa's proud declaration of mestizaje, the mixed-race heritage of Mexican Americans and their cultural productions. Describing the heritage of her subject, the Texas-Mexican vaquero, she says, "On one side, he descends from the first Americans, the Indians; on the other, his ancestry can be traced to the Spanish adventurer and conquistador. From the mingling of these two races a unique type has resulted, possessing not only salient racial characteristics of both but also certain peculiar traits created by the natural environment and surroundings in which he lives" (7). González was at the time, if not the only, then one of the few members of the Texas Folklore Society who was a Mexican American woman. Nevertheless, she dared to make it absolutely clear that Mexican Americans had Indian as well as Spanish blood running in their veins. That was a bold move in times when pseudo-identifiers, such as "Spanish," "Spanish American," "Latin," and "Latin American"—designed precisely to excise American Indian heritage—were the norm. Note also González's emphasis on the coevolution of the place and its peoples; she implicitly argues for the spirit that animates both and rejects coloniality's suspicion that "nature" and "natives" have no soul.

González's introduction establishes from the outset two themes that she sustains in the tales that follow: cultural hybridity and the human-nature connection. Few scholars have considered these first-published tales in depth.[6] The cultural hybridity and human-nature connection that González foregrounds in her introductory comments play key roles in the tales and are the venues by which she sounds her critique—especially in the story of how the mockingbird got its white wing feathers. Each tale appears under subheadings, each more or less independently of the others: "El Cenizo" (Texas Sage), "The

Mocking Bird," "El Cardo Santo" (The Thistle), "The Guadalupana Vine," and "Legends of Ghosts and Treasures." The story of "The Mocking Bird" comes second in the series and recounts the tale of how the mockingbird learned humility after a nasty run-in with a hawk and his ensuing rescue by a dove. The dove patches the mockingbird's wings with her own white feathers and the mockingbird concludes that all things happen by the grace of God. His white feathers are a permanent reminder of this episode. Here is the brief tale in full:

"The Mocking Bird"

> An equally interesting story tells how the mocking bird [*sic*] got the white feathers of its wings. There was a time when all the creatures of Nature talked a common language. The language was Spanish. *El zenzontle*, the mocking bird, had the sweetest voice of all. The other birds stopped their flight to listen to him; the Indian·lover ceased his words of love; even the talkative arroyo hushed. He foretold the spring, and when the days grew short and his song was no longer heard, the north winds came. Although he was not a foolish bird, *el zenzontle* was getting conceited.
>
> "I am great, indeed," he said to his mate. "All Nature obeys me. When I sing, the blossoms hid in the trees come forth; the prairie flowers put on their gayest garments at my call and the birds begin to mate; even man, the all wise, heeds my voice and dances with joy, for the happy season draws near."
>
> "Hush, you are foolish and conceited like all men," replied his wife. "They listen and wait for the voice of God, and when He calls, even you sing."
>
> He did not answer his wife, for you must remember he was not so foolish after all, but in his heart he knew that he was right.
>
> That night after kissing his wife goodnight, he said to her, "Tomorrow I will give a concert to the flowers, and you shall see them sway and dance when they hear me."
>
> "*Con el favor de Dios*," she replied. ("If God wills it.")
>
> "Whether God wills it or not I shall sing," he replied angrily. "Have I not told you that the flowers obey me and not God?"
>
> Early next morning *el zenzontle* could be seen perched on the highest limb of a huisache. He cleared his throat, coughed, and opened his bill to sing, but no sound came. For down with the force of a cyclone swooped a hawk and grabbed with his steel-like claws the slender body of the singer.

> "*Con el favor de Dios, con el favor de Dios,*" he cried in distress, while he thought of his wise little wife. As he was being carried up in the air, he realized his foolishness and repented of it, and said, "O God, it is you who make the flowers bloom and the birds sing, not I." As he thought thus, he felt himself slipping and falling, falling, falling. He fell on a ploughed field, and what a fall it was. A white dove who had her nest near by picked him up and comforted him.
>
> "My wings," he mourned, looking at them, "how tattered and torn they look! Whatever shall I tell my wife?"
>
> The dove took pity on him, and plucking three of her white feathers, mended his wings.
>
> As a reminder of his foolish pride, the mocking bird to this day has the white feathers of the dove. And it is said by those who know that he never begins to sing without saying, "*con el favor de Dios.*" ("Folklore" 10–11)

This charming story contains many artful arguments. José Limón has argued that, as many of these folktales appear again in her novel, *Dew on the Thorn*, the redacted stories "are artistically implicated in a running political commentary on ethnic, gender, and class relations" ("Introduction" xxi).[7] The stories, told by various characters, are woven into the narrative and move the novel's plot forward. Cotera has commented that González's "writings, from the start to the end, provided arguments against scientific and popular discourses, which had sought to describe, contain, and dispossess her people" ("Jovita González Mireles" 168). It is upon these points that I build: both on Limón's observation that the folktales figure into ethnic, gender, and class critiques, and Cotera's idea that González's work "provided arguments against scientific and popular discourses" that dispossessed Mexican Americans during the nineteenth and twentieth centuries. Of course, on the surface, this is a plain and simple morality tale, where a hubristic bird learns humility. It is also gently feminist, with the patient wife emerging as the hero at the end, possessing more wisdom than does the brash husband by providing him with the magic line that would save his life, "Con el favor de Dios."

But there is more here, too. Take, for example, an early line in the mockingbird story: "There was a time when all the creatures of Nature talked a common language" (10). This is a widespread, if not universal, element in American Indian origin myths. But then immediately we read: "The language was Spanish." Here, she makes an outlandish claim for the ubiquity of Spanish as the ur-language, not only of all human beings but of all other natural entities. This is not so outlandish if one considers the common spirit shared between human and nature. One can also see how she includes such a claim to show how Spanish was the language of the Rio Grande Valley before the American takeovers of 1836 and 1848, as well as before the agricultural transformation of the land in the early twentieth century. By juxtaposing a commonplace of American Indian mythology with an assertion of the primacy of the Spanish language, she reveals, in hardly more than a line of prose, the complications and contradictions, as well as pride, in the mestizo identity. Article VIII of the Treaty of Guadalupe Hidalgo guarantees Mexicans residing in territories acquired after 1848 a protection of all their rights as citizens, including the preservation of Spanish as their primary language, but González must have seen the abrogation of this right during the Jim Crow era in Texas. Also, with the use of the Spanish name for mockingbird, "el zenzontle," she clearly juxtaposes a melodic word, "zenzontle," alongside a significantly less flattering English name, "mockingbird." But the Spanish word "zenzontle" was borrowed from the Aztec language, Nahuatl, and means "four hundred voices" (Karttunen 32).

Thus, right from the beginning, González establishes themes of cultural hybridity by emphasizing aspects of Mexican and indigenous cultures still influential in South Texas and human-nature connection by recording a story that communicates a close observation of and sense of commonality with animals. These two elements gain momentum as the tale unfolds. By repeatedly using this "original" name for the mockingbird, González not only venerates the Spanish and Nahuatl languages, but she also drives home that linguistic and cultural hybridity. Moreover, in addition to the magic phrase "Con el favor de Dios," she includes two more words that, in this case, English has borrowed from the Spanish language, both denoting aspects of the natural environment. "Arroyo," a word then well integrated into regional English conversation, comes from the Spanish language and names a small, running stream. "Huisache" was also originally a Nahuatl word that Mexicans adopted into Spanish, with Anglo-Americans following suit, to name a shrubby acacia tree with fragrant bulbs that grows in Texas, especially in the Rio Grande Valley. With the integration of these terms, she clearly shows how the traditional language and knowledge of the natural environment endure.

The early notes of Mexican hybridity resonate with the story's resolution. For what else is the dove's blending of white feathers with the mockingbird's decidedly darker tones but an instance of racial mixing and hybrid creation? Moreover, her choice of narrative structure demonstrates a blend between two worldviews, a strategy that is all too familiar for mestizos. Native American tales often involve the communication of a lesson for humans told with animal stand-ins. Concerning Ojibwe tales, scholars J. Baird Callicott and Michael P. Nelson observe, "Animals in these stories are always personified and assumed to be endowed with such capacities as volition, reason, speech, and social interaction" (102). But the recognition of God's power over human fate—"el favor de Dios"—also signals the deep impression Spanish missionaries left in the Americas. The story's style and characters echo the narrative structure of Native American tales, while its message of humility before "Dios" is unquestionably inspired by Christianity. With these repeated references to hybridity and mestizaje, González makes a distinctly Mexican, and Mexican American, claim to the mockingbird.

It must also be noted that this story was published and read aloud at a meeting of the Texas Folklore Society, during the same year that Texas adopted the mockingbird as the state bird, 1927. How did the mockingbird's mestizo background sound to the ears of a largely Anglo-American and staunchly Texan audience? The folklorists appreciated González's work as a collection of quaint tales that perhaps did not threaten a mainstream understanding of history on the border. However, González's simple documentation certainly supplements the mainstream history with a deeper sense of the varieties of narratives in circulation at the time. She was, humbly enough, preserving the traditional environmental knowledge of the South Texas ranchers' way of life—with its attention to the seasonal changes, the drought cycles, and behavior of the plants and animals. Yet she was also positioning these folktales as responses to, and stern critiques of, the ecological revolution of her lifetime in the Rio Grande Valley. Cotera details this shift by citing the arrival of the railroad to South Texas in 1904 and the ensuing impact: "Within fifteen years of the construction of the railway system, the Texas Mexican people of the border region, with a few exceptions, had lost the world that Jovita González knew, 'the world of cattle *hacendados* and *vaqueros*,' and would come to live in 'a world of commercial farmers and migrant laborers'" ("Jovita González Mireles" 161).

The framework of ecological revolution offers us yet another way of making sense of these folktales. How, exactly, does this story specifically respond to this ecological revolution in the Rio Grande Valley? The mockingbird's lesson of humility offers a critique of the transformation of South Texas lands, water, and culture brought on by growing Anglo-American influence and land-use practices. The mockingbird learns that he cannot truly make the spring arrive, the flowers bloom, the birds sing. His early insistence on this power draws his wife's cautious remonstration, and almost ends in fatal disaster for him. But he manages to recover, acknowledging a higher and wiser power, and acquires some hybrid elements along the way. In the South Texas of the early twentieth century, droves of farmers were arriving, answering the call of advertisements spread across the Midwest and the South. The ads told them that a fertile oasis awaited them in the Rio Grande Valley. All they had to do was find some unused land, usually part of a rancho, and take advantage of the many consecutive days of sunshine and balmy weather, but upon arrival, they understood that they would need to irrigate. In a landscape that had heretofore been ranching territory, irrigation was not common. The newly arrived farmers, largely attracted by cheap land, sunshine, and the new railroad that could deliver their crops to the world, found a way to irrigate their crops. They used the waters of the Rio Grande, and so initiated the process of making spring arrive, just as el zenzontle vainly claims he can do.

However, the diversion of the Rio Grande initiated by the new farming practices started a process of water depletion in the river such that now the river struggles to make its way to the Gulf of Mexico, sometimes even trickling to a standstill before reaching its destination. In one simple tale, González laments the practices of the newly arrived farmers and critiques the way they participated in the decline of traditional Texas-Mexican ranch culture by hubristically instituting their neverending "spring" of irrigation fed by the Rio Grande. Her story begins with el zenzontle claiming to make the *prairie* flowers bloom, as they once did on the ranchos, and ends with the mockingbird falling "on a ploughed field," indicating the already transforming land. Nevertheless, she also indicates his acknowledgement of a higher power's control over the seasons: "O God, it is you who make the flowers bloom and the birds sing, not I" ("Folklore" 11). However, the transition to farming continued in the valley, and the acquisition of ranch land for farming purposes was not always through honorable means. According to historian David Montejano:

> The impact of the farm developments on the Texas Mexican people was profound . . . Taxes, mortgage debts, legal battles, the effects of the erratic cattle and sheep market, outright coercion and fraud, as well as the cash offer of land speculators, all combined once more to reduce the number of Mexican landowners . . .

> the primary sources leave little doubt on this point—the dispossession of landed Mexicans was a sweeping one. (113)

However, public opinion in the early twentieth century argued that simple market forces compelled the shift. A couple of Texas geographers made this case in the pages of the journal *Economic Geography* during the late 1920s and early 1930s. Their articles were concerned with the ranch-to-agriculture shift but in a tone—certainly reflective of the time—that celebrated the "progress" that farming brought to South Texas. In a 1932 article on "Land Utilization in the Lower Rio Grande Valley of Texas," Southern Methodist University geography professor Edwin J. Foscue reduces the transition in land ownership to a simple tale of economics: "Lands that are today in [*sic*] irrigated farms were once the widespread pastures of the Texas-Mexican rancheros, who held clear title to the land, and refused to sell, until modern development caused the land to become so valuable that they felt forced to sell" (3). In the face of such half-truths that populated academic circles in her day, González takes heart by documenting the longstanding presence of Mexican Americans in Texas and by communicating the environmental knowledge in the tales of the "Texas-Mexican vaquero," even as she was witnessing "one of the most phenomenal land movements in the history of the United States" (Montejano 108).

Another tale in her 1927 article also provides an artfully delivered critique of the newly arrived farmers' takeover of Texas-Mexican land, as well as further contextualizing the centerpiece mockingbird tale. The story about the origin of "El Cenizo," or the Texas sage tree, tells how this native Texas bush was a "gift of the Virgin" to the "cowmen" who had "gathered together and reverently knelt on the plain to beg for help" after a long drought (González, "Folklore" 9). The day after the people had prayed, they awoke to rainfall, and "as far as the eye could see, the plain was covered with silvery shrubs, sparkling with raindrops and covered with flowers, pink, lavender, and white" (10). They named the bush *cenizo* (ash) because the day was Ash Wednesday. Considered alongside the tale of the mockingbird's foolish pride, this example of the Texas-Mexicans' respect for a higher order's influence over rainfall shows the contradistinction between the farmers and the ranchers. Moreover, the tale concerns the Texas-Mexicans' integration of the cenizo, a native species, into a treasured cultural tale and further indicates their intimacy with the land and its ecological rhythms, even if such respect sometimes landed them in desperate situations. However, the arrival of farm agriculture threatened Texas-Mexican ranchers more than any drought. Foscue, one of the 1930s Texas geographers, documented the transitions for the ranchers in South Texas: "Cattle ranching which had been supreme in the area for more than 150 years was suddenly dislodged from its monopoly upon river lands, and driven into the back country, or out on the coastal prairie that was too wet to cultivate" (3). At the same time, these geographers also documented the ranchers' coexistence with native species, here described as the last remnants of the once dominant ranching culture:

> Areas unsuited to [farming] uses have been avoided by agricultural settlers and are chiefly useful for cattle raising. Extensive areas in the northern part of the region, coastal flats bordering the Gulf of Mexico, and marshy depressions between intermittent distributaries of the Rio Grande are of this sort. Much of this land will be cleared and drained ultimately for use in farming; but at present it is occupied by native grasses and thorny shrubs, and is suitable only for use as pasture. (Chambers 371)

The transformation of the land continued and dominated most of the valley, and in this 1930 article, geographer William T. Chambers recommends enterprise in South Texas due to its booming agricultural growth, its temperate weather, its "picturesque Spanish architecture" and its population: "The population consists of prosperous Americans and many Mexicans, the latter constituting a cheap and efficient labor supply" (373). With such rhetoric circulating about her native South Texas, one cannot deny the skill and passion with which González documents her community's experience. As early as 1927, González was making sure that this transition was recorded by means of the vaquero folktales. Gonzalez's writing shows how reciprocal relations with the local environment lead to a better and more stable sense of identity in this period. However, by 1964, when the next author in this chapter publishes his short story, "Man Without a Name," the Anglo-American establishment's interference in the relationship between Mexican Americans and the land created such confusion and destruction that the Chicana/o civil rights era rose up to create changes in the power structures to give Chicanas/os access to self-determination and freedom from oppression. Sabine Ulibarrí's story sets the tone for the crisis of being and of materiality that leads to outright political action.

CONCLUSION: SABINE ULIBARRÍ AND THE NEXT ERA OF CHICANA/O POLITICS

Sabine Ulibarrí, from northern New Mexico, wrote a story of the Mexican American struggle with identity and power: should one embrace or reject whiteness? Is the possession of land and access to power at the cost of integrity about one's identity worthwhile? This story indicates the psychic toll of denying indigeneity, and thus serves as a transition from the ambivalent identities in the early twentieth century to the strong identity claimed during the movement. Titled "Man Without a Name," the story concerns a young man who has written a book about his father. His father, the patriarch of their northern New Mexico clan, the Turriagas, is recently deceased and the son's attempt to write a book about him is much anticipated by the family.

On the night when the family gathers to celebrate his book, the young man finds himself haunted by the ghost of his deceased father. Presumably, his father visits him because the book the young man wrote critiques his father's life rather than just celebrates it. When he reads portions aloud to his family, they "began to perceive the reality behind outward appearances" (8). The ghost does more and more to drive the young man crazy, convincing him that he is no longer himself but his father reincarnate. The family helps in this process: "They had decided that I was no longer I! That I was my father!" (14). The centrality of identity and the impact upon it from external forces is undeniable in this story.

One important scene in the haunting of the young man by his father occurs on the family's ancestral land. His father represents an identity that affirmed whiteness and denied indigeneity as a strategy to retain claim to their land grant. However, that land is now "violated": it is fenced off, owned by someone else, and it even has a "no trespassing" sign installed. The young man goes there to fish, trespassing despite the sign, and tries to forget that his father inhabits his every thought. However, his father continues to haunt him, telling him where to go and where to cast his line to catch a big fish. The young man tries to resist his father's commands, determined that he would "go his own independent way" (42). However, his "own independent" fishing strategies get him nowhere, and despite himself, he casts his line exactly where his father had been insisting: next to a big rock, in a spot of sunshine. He immediately catches a big fish, and even unwittingly yelps for joy. Yet his joy is short-lived, as he realizes that his success is only due to his father's instruction: "How bitter is the taste of humiliation! In the excitement of gaining a bagatelle, I had lost a soul, I had bartered my integrity, I had allowed myself to be manipulated by alien forces—and for a moment I had enjoyed my own defeat!" (44). He throws the fish back into the water, refusing to profit from the ghost's advice, and he feels humiliated, especially in the presence of the fish whose "white belly, undulating on the water, accused" him (44).

Ulibarrí published this story in 1964, in contrast to the earlier dates of the other works in this chapter, and therefore Ulibarri has more distance from the Spanish American identity. Yet he still endeavors in his writings to document the life and culture of his northern New Mexican family in the face of a growing trend of assimilation. His young protagonist, who remains the "man without a name," must choose whether to inherit his father's identity whole, as the story suggests, or if he will persevere with his plan to "go his own independent way." However, he is in perhaps a worse position than was his father: at least his father still owned land when he was young. Now a member of the next generation, the young man must contend with finding his own identity in a condition of separation from his ancestral landscape (the young man lives in Albuquerque). He also feels he must reject the privilege of whiteness (as represented by the white belly of the big fish). By the end of the story, we see him completely consumed with confusion and alienation, yet with his wife by his side as a hope for a future recovery of himself. The ambivalent ending challenges readers to struggle with the man's identity: should he work to reclaim himself even if he has no clear direction, or should he give in to his father's image and therefore gain the privilege of whiteness but lose his independence and his integrity?

Considering the coloniality of being in relation to these five writers gives Chicana/o literary history a new way to appreciate and value this period of writing. This analytic frees Chicana/o criticism from consistently evaluating our works only in terms of explicit social justice agendas that work to reform or revolutionize ideological or institutional structures. Decolonizing the human spirit and its relation with the natural environment ranks just as high as these agendas.

3

"LA SANTA TIERRA"

Chicana/o Writers Transcending Possession in the Late Twentieth Century

On Sunday, June 8, 1969, bright orange flames licked at a wooden sign reading "Santa Fe National Forest." Not the result of an accidental forest fire, this sign's destruction was intentional: to show a community's disgust with the Forest Service's management of lands taken from them by the U.S. government 121 years prior, after the conclusion of the U.S.-Mexican War in 1848.[1] In the 1960s, the people of northern New Mexico made known their acute awareness of Mexican Americans' threatened intimacy with the land. Their burning of the sign signaled their particular alienation from the idea of state-managed nature exemplified by the national forests. Of all places, why did national forests emerge as targets? Northern New Mexico families maintain a strong link to their past, and in the 1960s this meant a keen awareness of the lands they had lost to government takeover and private enterprise due to corrupt and compromised processes of land-grant reviews in the late nineteenth century.[2]

Scholar Jake Kosek states that "land politics in northern New Mexico are not simply struggles over narrowly defined ideas of property but instead are invested with regional histories of racism, national exclusion, and state violence" (189). The Mexican American population in New Mexico had too long endured federal management of lands they always perceived as historically being in their care. Combine this perception with a growing difficulty in surviving via the land and an increased awareness of civil rights organizing and you get a community ready to fight. The Forest Service kept the public, including Mexican Americans, from grazing their sheep and gathering their firewood on public lands. These policies instigated animosity from the start and earned the Forest Service a derogatory nickname within the Mexican American community: "La Floresta."

Mexican American activists targeted national forest lands by setting fire to two Forest Service signs in 1969. This was not the only antagonism on national forest lands. In 1966 the same group of activists occupied an amphitheater. For five days in October 1966, approximately three hundred members of a Mexican American organization, La Alianza Federal de Mercedes (The Federal Alliance of Land Grants), took over the Kit Carson National Forest lands surrounding the Echo Amphitheater located near Abiquiu, a place most known for inspiring Georgia O'Keefe's striking landscape paintings so consistently devoid of human figures. La Alianza, in honor of the longstanding Mexican American inhabitants of the area, gave the seized territory a new name: Republic of San Joaquín del Río de Chama. They derived this name from the Spanish-era grant that declared the area property of Spanish/Mexican settlers—a grant that was not upheld in the land commission hearings following the U.S.-Mexican War.[3]

La Alianza worked toward recovering such lost lands in northern New Mexico during the 1960s, with some of these land disputes continuing in the courts today.[4] In the Kit Carson National Forest during those five days in 1966, La Alianza elected officials for their new nation. When forest rangers tried to remove the group from federal property, the newly elected officials captured the rangers and put them on trial in their new nation. The activists soon after released the rangers, but not before making La Alianza's sovereignty known, although its sovereignty was fleeting. The actions of La Alianza in New Mexico were publicized across the nation, reaching the pages of the *Washington Post* and other major news outlets. Led by the charismatic former evangelical preacher Reies Lopez Tijerina, it is no wonder their group caught the spotlight. Their activism also dominates popular knowledge of the Chicana/o history of this period. Literary critic Randy Ontiveros observes that La Alianza "played an important role in the formation of Chicano/a environmentalism" (100). Ontiveros also states, "In contrast with mainstream environmentalism, La Alianza did not fetishize the New Mexican landscape[;] . . . they used their cultural heritage as a resource for confronting state and corporate power and for building a grassroots environmental justice movement" (100). While it is certainly important and, indeed, vital to be aware of La Alianza's strength and courage, their dominance in

Chicana/o historiography thus far has kept other voices and expressions from being heard. There is, after all, another story here. The genealogy of Mexican American goodlife writing makes space for these other voices to express less a desire to possess territory than a commitment to dwelling with the land and thus defying the hierarchies of modernity/coloniality.

In his study on this historical period, *In the Spirit of a New People: The Cultural Politics of the Chicano Movement*, Ontiveros dedicates a complete chapter to "Green Aztlán," especially noting the environmental aspects in visual art during this dynamic era. Ontiveros argues that in light of the Chicano/a community's "growing [concern] about the impact of industrialization on the natural world," their "response was to formulate a distinctive Chicano/a environmentalism, one that drew on Mexican American traditions and that joined ecological concerns with social concerns" (88). Ontiveros also aptly asserts that "little attention has been paid to the complex relationship between the Chicano civil rights movement and the modern environmentalist movement, even though they appeared on the public stage simultaneously" (87). I certainly agree with Ontiveros about the curious absence of analysis about the simultaneous emergence of these two important social and political movements, but I also want to emphasize how Chicana/o writing during this period explores environmental issues in a way that connects with an already established literary history concerned with the human-nature relationship. In particular, this chapter will engage with writings concurrent with the land rights movement that have not received as much critical attention and that emphasize a nonpossessive relation with the land.

The writers in this chapter make visible a quiet but powerful and less recognized aspect of the Chicana/o civil rights era (1960s–1970s), which appears more explicitly in the 1980s and 1990s, especially in Chicana feminist writings. These writers show that not all the activists during the civil rights era made direct demands on the existing institutions, such as government, courts, schools, and employers. They saw these institutions as already corrupt and not worth attempting to reform. Enacting typical decolonial strategy, they sidestep the mainstream structure of power by constructing their own system of values consonant with the genealogy of the goodlife that I trace in this literary history. They did not all seek the repossession of lost lands, nor did they appeal to the courts to reunite them with the land. Instead, they sought to create a new way of relating to the land and to one another that had nothing to do with ownership. They sought to build their own institutions and support networks.

They rejected the modern/colonial ideas of ownership and sought to transcend possession.

In this chapter I discuss a number of works from the last half of the twentieth century. First, I examine writings from a newspaper published during the late 1960s and early 1970s called *El Grito del Norte* in northern New Mexico. In this newspaper, Chicana feminist writers explain that the possessiveness of coloniality is ultimately to blame for the threat to Chicana/o culture and that the people should resist by transcending the idea of possession itself. Then I examine a classic novel from 1972, *Bless Me, Ultima* by Rudolfo Anaya, which heretofore has seemed anomalous in its exploration of 1940s Mexican American *curandera* culture. In the fiery context of 1970s Chicano activism, Anaya's novel was appreciated for its beauty, but its quiet strength was little noticed and sometimes even misunderstood. However, in the context of goodlife writing, it appears completely in tune with the efforts to establish and maintain the unique Mexican American relationship with the natural environment. Next, I offer a reading of a collection of oral histories recorded in the 1980s but that span the length of the twentieth century. The women in this collection describe modernity's abstraction of the natural environment and their families' ensuing alienation from the land. Finally, I conclude with a reading of Gloria Anzaldúa's *Borderlands/La Frontera*, which links her Chicana feminist voice to the lesser-known women writers and thinkers discussed earlier in the chapter. All of these works emphasize the community ethic that drives this culture and their ongoing relation with the land.

ELIZABETH "BETITA" MARTINEZ AND ENRIQUETA VASQUEZ IN *EL GRITO DEL NORTE* (1968–1973): SMOKEY BEARS AND HEALING SERPENTS

El Grito del Norte was a community newspaper published from 1968 to 1973 in Española, New Mexico, and dedicated, "To advance the cause of justice for poor people and preserve the rich cultural heritage of la Raza [the Chicano people] in this area." Elizabeth "Betita" Martinez and Beverly Axelrod founded the newspaper in 1968 in support of La Alianza's activities. However, the feminist structure and writers of this newspaper pushed the agenda beyond the land ownership goals of La Alianza. The feminist voices in *El Grito* challenge the community to embrace their history of working with the local environment,

FIGURE 2. Smokey the Bear from *El Grito del Norte*.

and in so doing, enact an ecological cooperation that ultimately means more than land titles or prices per acre.

In a lengthy article, illustrated by the little-known aggressive version of Smokey the Bear (see figure 2), Martinez details the history of Mexican American land dispossession, dating back to Spanish colonization. In a quote that reveals the tension between the Forest Service and Mexican Americans in New Mexico, Martinez says:

> El peor golpe que recibió la Raza fue en 1906 cuando el Presidente Theodore Roosevelt estableció el sistema de parques nacionales con millones de hectares de los terrenos que pertenecía al pueblo. En años recientes la Floresta tiene como lema a "Smokey el Oso," que a muchos norteamericanos parece amable, pero es de odio para el indo-hispano. Smokey es símbolo del administrador de la Floresta, por lo regular un Anglo, que favorece a los grandes rancheros comerciales en contra de los intereses de los hispanos con el poco ganado que les queda. (13)
>
> [The worst insult that la Raza experienced happened in 1906, when President Theodore Roosevelt established the national park system with millions of hectares of the lands that belonged to the [Chicano] people. In recent years, the Forest Service [as noted, derogatively known as "La Floresta"] has as a mascot Smokey the Bear, which to many Americans seems loveable, but he is hateful for the Indo-Hispano. Smokey is the symbol of a Forest Service administrator, usually an Anglo-American, who favors the large commercial ranching operations over the interests of the Hispanos, who have very few livestock left. (my translation)]

Conventionally, within environmental studies, the creation of the national park system is an act of great wisdom, even dubbed "America's best idea" by acclaimed documentarian Ken Burns.[5] However, Mexican Americans have had a vexed relationship with national parks and the value of wilderness, especially when that means the absence of human habitation. It is no surprise, then, that anthropologist Jake Kosek found antagonism toward Smokey the Bear when he arrived in northern New Mexico to conduct research about the forests in the 1990s: "Racialized nationalist histories underlying the Smokey Bear campaign have infused the bear and the forest with exclusionary formations of U.S. nationalism and, together with the acquisition and misuse of forest lands by the Forest Service in northern New Mexico, have made Smokey more of a 'white racist pig' and a 'despotic land thief' than a benevolent protector of public forests" (186). The quotation from Martinez's article speaks to this dim view of the Forest Service mascot.

The territories that the United States conquered and acquired from Mexico in the nineteenth century made the U.S. national park system possible. Indeed, the monumental beauty of Alta California's Yosemite helped John Muir persuade Theodore Roosevelt to create the national parks. Yet, of course, without the U.S-Mexican War, Yosemite would not have been a U.S. territory. Martinez notes the way that the Forest Service limits local herdsmen's access to federal lands but encourages commercial ranchers to use the same lands on a larger scale. One might want to say that perhaps this was just a land-use oversight that the Forest Service rectified, but the issue of ethnic identity and discrimination cannot be ignored here. Martinez also claims that we all know that the lands of the Southwest "belong first to the natural Indian of the North American continent and then to the Indo-Hispano . . . And we know that the discrimination against us is because of the color of our skin, our language, and our culture (our system of values)" (12, my translation). Mexican American identity encompasses centuries of living on the North American continent, including a negotiation of hybrid identity: Spanish/European, Indigenous/Native American, and also African and Asian—the way Martinez puts it is "Indo-Hispano."

Mexican Americans do not always successfully negotiate our hybrid identity, but the process is instructive. As we become a planet of increasingly hybrid identities, the Mexican American experience can teach us a great deal. Enriqueta Vasquez also wrote a regular column for *El Grito* titled "¡Despierten Hermanos!" (Awaken, Brothers and Sisters!) from the year 1968 to 1972. In one article she details the hybrid Mexican American identity: "La Raza is a mestizo nation. Our Spanish blood predates 300 years in Las Americas; our Indian blood predates 25,000 years and we are the mestizo nation of Aztlán, the nation of mixed bloods. In realizing this, we come home to our true selves, to our indigenous being, to our Indian family ties . . . We choose; we go the way of the land; the way of the earth; the way of the water; the way of the wind; the way of AZTLÁN" (56).

In the above-quoted column, titled "La Santa Tierra," Vasquez argues that land possession is a foreign concept for Chicanas/os. The Chicano Civil Rights Movement in the 1960s capitalized on the proximity of Mexican Americans to the natural environment through their work places and poor living conditions. Rather than feeling ashamed of their social positions and manual labor employments, Mexican Americans took pride in their contributions and

FIGURE 3. La Serpientita from *El Grito del Norte.*

demanded recognition for such. Vasquez's columns in *El Grito del Norte* offer a snapshot of these commitments, arguing that living with nature means not possessing it: "We can well understand why to a people who are capable of living in harmony with nature, it is very difficult to understand the concept of ownership and possession of land" (55–56). Here Vasquez details a good-life approach that affirms the agency and sovereignty of nature. Ontiveros also notes how Vasquez "wanted to make issues of ecology a part of the Chicano movement" (98).

Two more examples of articles from *El Grito* illustrate the varied ways that these writers communicate the agency of nature: one is a folktale intended for children and the other is a traditional remedy for skin sores told via a personal story. The first is titled "La Serpientita" (The Little Serpent). What follows is a brief summary of the tale.

A little girl finds a tiny snake while playing outside and decides to take care of it, feeding it, keeping it in a bottle until it grows too large, then keeping it in a box, until it outgrows that as well. Finally, the snake tells the girl that it is too big for the girl's care and she must let it roam free. The little girl dearly loves her little serpent and she does not want to let it go, but she finally does let it go free. The serpent tells her, "If you ever need me, just call me." Time passes and the girl grows into a beautiful young woman. Other women in the village are jealous of her beauty and goodness. They maliciously tell the king that the girl claims she can turn anything she touches into silver and gold. The king summons her. She fails the test, and he sends his guards to pluck out her eyes and dump her in the wilderness. Once alone, the girl calls the little serpent, who immediately appears. The serpent asks how it can help, and the girl tells it everything that happened. The serpent licks the girl's hands and says she can now turn objects into gold and silver. The serpent then leaves, and shortly after, a firewood gatherer appears, a little old man and his mule. The girl sends the firewood gatherer into town to buy two eyes with two nuggets—one gold and one silver. The women who lied to the king are greedy, and they have kept the girl's eyes to remind them of their deed, so they exchange them with the firewood gatherer for the gold and silver nuggets. When the firewood gatherer delivers her eyes back to the girl, her vision is restored and she returns home. After some time, she hears that the king has sent hunters to kill a gigantic serpent in the forest, so she runs to warn the snake and it escapes. The girl returns home, where she lives happily ever after, except that at times, she misses the company of the snake.

So much can be said about this folktale. It depicts petty relations among women—the jealous ones who lie to the king because of the girl's beauty. Yet the story transcends the human-to-human relations to make a bigger point: the girl survives only because the snake helps her, and the snake only survives with the girl's assistance. Both are threatened by human emotions: envy on the part of the village women and greed on the part of the king, who wants a subject who can turn things into precious metals. A lot can be read in what is not said here as well: the girl uses her special powers only once and only to regain her sight. Otherwise, she doesn't care to turn things into gold and silver, while gold and silver during colonial times drove colonizers' exploitation of indigenous peoples throughout the Americas. Far from the Christian tradition of demonizing reptiles, this story (a children's story, no less!) offers a keen awareness of the interdependence of humans and the natural environment. The girl/woman and the snake exist in mutual respect, with no need to claim one another as a possession or in any permanent defined role other than as someone to call upon if the need arises.

Another item, titled "La Voz de Nuestra Cultura" (The Voice of Our Culture), appeared periodically as a column on herbal remedies during the run of *El Grito*. Already in the title, the author gives agency to the culture, which speaks to and teaches the people. This column is dedicated to detailing the healing power of snake meat and snake powder. It tells of a man in Michoacán who suffered so badly from skin sores that he thought he would die. He goes to a remote cave to live his final days in solitude, eating what he gathers in the local area and drinking from a tin pail that he dips in the local stream. As time goes by, his health improves. Finally he gets so much better that he decides to return to his family. He picks up his drinking pail, dumps out the water, and at the bottom he finds the remnants of a snakeskin. The column explains that it is this snakeskin that helped him recover from his illness. Readers are free to take this story for its literal value—use a snakeskin to help heal skin sores. Readers are also free to consider this story somewhat apocryphal, but, even in its whimsy, this story calls upon a natural remedy to heal the man's serious illness. The story creates a parallel between the snake's sacrifice of her skin to provide for the restoration of the man's skin, and in so doing offers an example of a reciprocal relationship between human and nature.

Gloria Anzaldúa explains the value of snakes in indigenous and Chicano culture in her book *Borderlands/La Frontera*. In "Entering into the Serpent," she associates the Earth with a snake: "Like the ancient Olmecs, I know Earth is a coiled Serpent. Forty years it's taken me to enter into the Serpent, to acknowledge that I have a body, that I am a body and to assimilate the animal body, the animal soul" (48).

She associates the image of a serpent with the Earth itself—the natural environment. But she does not stop there. She also extends the metaphor to herself—she finds her bodily materiality in the act of "enter[ing] into the Serpent" and in so doing she realizes her direct connection to the animal world and the greater natural environment. These few examples from *El Grito* show how an alternative, nonpossessive approach existed in the Chicana/o civil rights era.

RUDOLFO ANAYA'S *BLESS ME, ULTIMA* (1972): DECOLONIZING SPIRITUALITY

Bless Me, Ultima, one of the most celebrated Chicana/o novels, complements the women's voices in *El Grito del Norte* in several ways. Written in New Mexico, in the same historical, geographic, and political context as the newspaper, *Bless Me, Ultima* also shies away from any overt political confrontation between Mexican American and Anglo-American cultures. As scholar Horst Tonn observes, "The novel does not seem to harmonize with the historical context, the militant and disruptive political climate of the late 1960s and early 1970s" (59). Scholar Alex Hunt also remarks that "some critics . . . regard it as a work of escapist nostalgia when the times called for social protest" (180). Anaya instead focuses on tensions within the Mexican American community in the 1940s. The spotlight on a rural agricultural community also gives readers a window to New Mexican traditional environmental knowledge. This novel's celebration of the curandera Ultima and the practices she preserves through her mentorship of Antonio Márez shows this book's stealthy resistance. *Bless Me, Ultima* argues on behalf of Ultima's hybrid spirituality in the face of colonial imposition of Roman Catholicism and shows how Chicana/o everyday practice reinforces a reciprocal relation between humans and nature, even granting nature its own subjectivity and autonomy.[6] Antonio attends school during the period of his life captured in this novel, but we rarely see him in class. Rather, readers see him learning in informal settings—from curandera Ultima, his religious mother, his vaquero father, his farmer uncles, the village priest, and his peers. This emphasis on noninstitutional learning, with the exception of the church, decolonizes Antonio's knowledge-acquisition process and deconstructs the hierarchy of epistemology. *Bless Me, Ultima* shows that the community can face the difficult challenges of gaining civil rights once they recognize and reinforce hybrid spirituality and the relation between humans and nature.

The centrality of agriculture to Anaya's text becomes obvious as the young protagonist Antonio (Tony) Márez struggles to decide which of his parents' traditions he will follow: his father's way of ranching or his maternal family's practice of farming. We meet Tony when he is on the brink of his seventh birthday during the summer before he begins attending school. Tony's youth allows readers to observe the development of a young boy whose home life still makes a great impact on him and whose identity also begins to shift as a result of his experiences at school. Wanting Tony to carry on his or her respective family's traditions, each parent competes to influence Tony, the youngest of five siblings. The difficulty for Tony lies in the fact that each of his parents' hopes for him differ so much.

Anaya presents his bildungsroman in a unique way. While some might compare Tony's journey to Stephen J. Dedalus in James Joyce's *A Portrait of the*

Artist as a Young Man, readers see Stephen largely in institutional settings. Tony roams the llano and follows the river in his quest for knowledge, all the while debating between his mother's and his father's desires for his future endeavors. His father explains why his vaquero tradition excels—in part because he did not need school as a boy, preferring instead his special relationship with the llano: "Me, my father gave me a saddle blanket and a wild pony when I was ten. There is your life, he said, and he pointed to the llano. So the llano was my school, it was my teacher, it was my first love—" (51). Throughout the novel, Tony's father repeats his belief that the life of a vaquero teaches everything a man needs to know and that the nomadic lifestyle offers the deepest satisfaction. Even his surname, Márez, means oceans and Tony speculates that the name describes "an exuberant, restless people, wandering across the ocean of the plain" (5). Still, the senior Márez married a woman from a farming family, the Lunas, and she desires her youngest child to emulate her family's quiet ways and honor their deep-rooted cultivation of the land by becoming a rural priest.

The contrast between tilling soil and herding animals in *Bless Me, Ultima* resonates with a familiar story from the Bible, the conflict between Cain and Abel. Cain is the firstborn son of Adam and Eve, and he chooses to farm fruits and vegetables, while his younger brother Abel opts for shepherding. Their vocations are complementary, and they only experience discord after God asks them to make him an offering. Cain lays his best fruits and vegetables on the altar, and Abel chooses a beautiful lamb for sacrifice. God expresses a preference for Abel's offering of flesh, which results in Cain harboring a deep jealousy toward his brother. Cain eventually murders Abel. When God inquires after Abel's whereabouts, Cain famously replies, "Am I my brother's keeper?" With this, God banishes Cain away to wander the world alone.

This story's warning against jealousy, especially within a family, connects with this novel's concern with the tension found among the members of this small New Mexico community. Tony gets along well with his siblings, so the Cain and Abel story's parallel indicates conflict that needs resolution among the adults who compete for Tony's attention. The lessons that Tony learns on the llano predate the arrival of large-scale agriculture and show readers how the preservation of cultural heritage goes together with protection and appreciation of the natural environment. Tony realizes that "the dreams of my mother were opposite the wishes of my father. She wanted a priest to watch over the farmers of the valley, he wanted a son to travel with him . . ." (69). Throughout the novel Tony wonders if he even possesses the abilities that could fulfill his mother's ambitions for him: he recognizes that priesthood is a very special calling. He longs to know his future, and this unknown drives the narrative: readers also want to know which path he will choose. Tony suspects that Ultima, the elderly healing woman who joins his family's home at the novel's beginning and whose name means "the last one," possesses this knowledge. He knows she was midwife at his birth, he shares a deep connection with her, and he even presumes that "she held the secret of [his] destiny" (11). Interestingly, it is he that holds the key to Ultima's life—if he takes up her healing ways, she need not be "the last one."

Tony's parents want him to choose between two different vocations, both intimately related to the Earth, but the arrival of Ultima gives Tony a third alternative that does not exist for the Bible's Cain and Abel and that Tony's parents do not even imagine possible for their son. Tony can choose an option outside the realm of Western Christianity and scholarship. Ultima's calling as a curandera seems to fit Tony's affinities, demeanor, and abilities. From the first paragraph of the novel, readers see Tony's world change as a result of Ultima's mentorship: "When she came the beauty of the llano unfolded before my eyes, and the gurgling waters of the river sang to the hum of the turning earth . . . She took my hand, and the silent, magic powers she possessed made beauty from the raw, sun-baked llano, the green river valley . . . My bare feet felt the throbbing earth and my body trembled with excitement" (1).

Ultima's medicinal practices bring Tony close to the natural environment, and he feels wisdom and peace as a result. He still longs to know his future vocation, but spending time with Ultima gives Tony a way to bring together the disparate worlds of his parents. When reflecting on the interdependence of the Earth, moon, and sun, Tony reasons, "The sun was good. The men of the llano were men of the sun. The men of the farms along the river were men of the moon. But we were all children of the white sun" (25). He begins to see a whole system with complementary parts rather than individual, different, and competing elements. The story of Cain and Abel appeared in the New World with the Christianity of the European colonizers, and the story details a separation between two modes of domestication that do not necessarily compete with one another. On the other hand, curanderismo dates back to pre-Columbian times, to the traditions and knowledge of many different indigenous tribes. The wisdom of curanderismo, folk healing, gives Tony a better way to understand his parents' backgrounds. Tony's alternative choice, the possibility to become a curandero, indicates the indigenous connection to the natural environment maintained by his people in northern New Mexico, even after centuries of influence from European Christianity. His preference for healing also rises above the dichotomy of gender roles available to him. He makes his choice to spend countless hours with Ultima in defiance of the expectation that he will opt for a masculine role as a vaquero, farmer, or priest.

Further, both ranching and farming involve the practice of domestication—one of animals, the other of plants. Both represent a fateful step on the part of humans all over the planet away from the more ancient human lifestyle of hunting and gathering—and thus of encountering the unmanipulated natural environment as it presents itself and foraging for a living. As a curandero, Tony can not only transcend the dichotomous choice his parents present to him, he can also transcend the impulse to transform nature; and he can thus return to a more complete partnership with Earth.

The story of the Golden Carp shows just how this novel resolves Tony's conflict by letting him connect to an indigenous relationship with his local environment that even predates the ancient competition between herdsmen and farmers. When Tony celebrates the end of his first year of school, he goes fishing with his friend Samuel, who takes the opportunity to tell Tony the reason why people of the village avoid catching and eating the carp in their local river. Tony observes the practice, but he has never known why the carp are prohibited. Samuel offers an explanation by way of an indigenous creation story, complete with "*the people*" wandering homeless until the gods let them dwell in the river valley (73). The gods give the people all they need and desire, with the exception of the carp in the river. The people abide by the gods' wishes until they suffer a forty-year drought. Finally, the people cannot resist catching and eating the carp, and the gods become angry. Before the gods follow through with their punishment of the people, one god pleads on their behalf and prevents their destruction. He pleads with the gods to punish the people only by turning them into carp. In return, he volunteers to become a carp alongside them. The gods relent and turn the people into carp. They also turn the god who intervenes on their behalf, but they make him larger and more beautiful with golden, shimmering scales in honor of his status. The merciful god becomes the Golden Carp. Critic Alex Hunt offers a detailed analysis of the ecological consciousness that the Golden Carp mythology represents in this and others of Anaya's novels. Hunt's tracing of the Golden Carp throughout Anaya's oeuvre shows the transnational connections of this Anaya-created mythology. Hunt also shows how deeply the Golden Carp mythology connects with Aztec mythology, especially creation stories.

The story thrills and alarms Tony. On one hand, he accepts the story intuitively—"I could not disbelieve Samuel" (Anaya 75). On the other, his years of Catholic indoctrination resist Samuel's story: "The roots of everything I had ever believed in seemed shaken . . . If the golden carp was a god, who was the man on the cross? The Virgin? Was my mother praying to the wrong God?" (75). Slowly, Tony's experiences repeatedly show him the Christian God's power overshadowed by the indigenous way of harnessing the power of nature. Anaya has revealed that he created the story of the Golden Carp himself, rather than basing it on a traditional indigenous story. However, he also notes that he drew upon the idiom of many different indigenous creation stories to create this one. What becomes most important about the Golden Carp for Tony is not necessarily its authenticity as much as the way it calls into question the Christianity that infused his imagination up until the point when Samuel offers him an alternative reality.

An example of the way Tony sees God's power diminishing in light of indigenous practices is through Ultima's rescue of his uncle from certain death. Her herbal remedies and rituals, dating back to indigenous methods, keep his uncle from dying. In another instance, an evil man accuses Ultima of killing his daughter, but her owl protects her by plucking out one of his eyes and keeps him from following through on his threats to unmask her as a witch. Moreover, only Ultima and the other boys who know the story of the carp feel the same presence from the river as Tony does, as if the river were a conscious entity. Even when Tony imitates the way Ultima walks, he gets closer to nature: "There was a nobility to her walk that lent a grace to the small figure. I watched her carefully and imitated her walk, and when I did I found that I was no longer lost in the enormous landscape of hills and sky. I was a very important part of the teeming life of the llano and the river" (37).

Looking at this novel with a goodlife writing lens brings out a new understanding of what Tony's choice of profession means in the long struggle for survival of indigenous cultures and practices. Despite his parents' best efforts, Tony discovers among his peers and with Ultima's assistance a possible life wherein he can resolve conflicts, heal his people's illnesses, and maintain a connection to his pre-Columbian, indigenous roots. Critic Marta Caminero-Santangelo argues that *Bless Me, Ultima* tells readers more about the suppression of indigenous identity within Mexican American communities of this time than it does about the continued connection to Amerindian roots. She aptly points out that not one of the Mexican American characters explicitly identifies as indigenous. However, this is more a reflection of politics in post–World War II America than it is an indictment of the people we come to know in Anaya's novel. What we witness are individuals and a community attempting to maintain a link to important knowledge and practices, even under the pressure of significant prejudice. Tony hears Ultima explain that her remedies emerge from a deep past: "She spoke to me of the common herbs and medicines we shared with the Indians of the Rio del Norte. She spoke of the ancient medicines of other tribes, the Aztecas, Mayas, and even of those in the old, old country, the Moors" (Anaya 39). More than simply a story about a boy growing up and choosing a profession, this novel symbolizes the recovery of an indigenous identity and the way this identity preserves a positive relationship with the natural environment. The oral histories in the next section also speak to strategies that negotiate this complex identity.

PATRICIA PRECIADO MARTIN'S *SONGS MY MOTHER SANG TO ME* (1992): LABOR AND THE SPIRIT OF THE LAND

Patricia Preciado Martin conducted oral histories with ten Mexican American women whose lives spanned the length of the twentieth century—from the 1910s to the 1990s—in southern Arizona. Preciado Martin fashioned her interviews with these ten women into standalone narratives and collected them in the book *Songs My Mother Sang to Me*, published in 1992. In so many ways, these are all tales of simple lives, hard work, family intimacy, and love of the land. But one need not scratch too far below the surface to hear more from *las voces de las mujeres* (the women's voices). With just a little attentiveness, we can hear their collective decolonial vision that criticizes the transformation that modernity enacted during their lifetimes—all from the seemingly simple recounted details of their lived experiences.

Critic Mary Pat Brady astutely observes the wisdom in one of these oral histories when she draws from it a fundamental concept for her book—the idea of "extinct lands." In the first oral history of Preciado Martin's book, Livia León Montiel explains that she wants to pass along to her descendants one small portion of her family's ranch land because so many of her family members live in so many different places now and she would like to provide them at least one stable point of reference. She says: "I'd like to leave this piece of land just as a gathering place for all of them . . . Land is becoming extinct, and all this will be developed. If it would only be possible for them to hold on to this little piece of land for all of them to reunite from time to time. This is my hope and my legacy." Brady's 2002 book *Extinct Lands, Temporal Geographies: Chicana Literature and the Urgency of Space* takes up more contemporary works—Sandra Cisneros, Gloria Anzaldúa, Terri de la Peña, Cherríe Moraga—but Brady takes this oral history as her point of departure to argue for the way this idea of extinct lands insightfully describes what is at stake during the transformation of the land and of Mexican American culture during the twentieth century. "Extinct lands," Brady suggests, "turn from lived, embodied space to the abstract space of capitalism" and she argues that Chicana literature attends to and urgently challenges this transformation (5).

I agree with Brady that Mexican American literature offers a much-needed window on these experiences, and I turn to these oral histories that tell us how Mexican Americans practiced a unique and everyday environmentalism, sensitive to the way coloniality attempted to intervene in their relationship with the land. Esperanza Montoya Padilla spoke with pride of her mother's skills as a curandera—"she had all kinds of little jars with herbs"—and as a *partera* (midwife), and of the way the locals "had faith in her healing" (Martin, *Songs* 107). She described her mother as so dedicated that she would "go wherever she was needed . . . She never complained" (107). Montoya Padilla also admits to not letting her mother help her with the birth or the care of her own children:

> I remember when I had my babies, I went by the book. Everything the book said, I did. And my mother would tell me to give the babies tea when they got sick, and I would say, "No, no, no! I don't want to start your medications!" And she would get upset with me and say, "How do you think that all of you were raised?" "Well, I'm going to bring mine up different," I would tell her. "I'm going to be the modern one!" (107)

Here we see Montoya Padilla admitting to the impact of "modern" ideas on her relationship with her mother and her lack of regard for her own upbringing and the cultural legacy her mother offers her. She realizes that she made a decision for her own children that kept them from living according to an organic way of being that she experienced in her own childhood. She rejects this tradition in favor of "the book," which, I would argue, is symbolic of Western epistemologies that claim precedence over much older and traditional ways of healing, orally transmitted. However, later in the interview, Montoya Padilla recognizes the value of her mother's knowledge and seemingly repents from her earlier faith in "the book": "My mother couldn't send me to school, but she gave me the greatest school in the world, how to be a good mother and take care of my children. She struggled so much, but she taught me the good things of life" (111). Here we see Montoya Padilla admitting that despite her faith in "the book," "the book" did not have much faith in her. She describes a situation in which her mother could not afford to send her to school, but of course many other factors that put the lauded "book" into a position of authority over her mother's traditional knowledge also colluded to keep young Montoya Padilla from going to school. Other oral histories in the collection describe children's impoverished experiences even when they did get to go to school—their teachers had little education themselves and often abandoned the students after just a couple of weeks in the harsh southern Arizona environment.

What we see here with Montoya Padilla is an example of her recognition of the valuable traditional ecological knowledge that her mother preserved, and attempted to teach her, and the way she arrives at her own decolonial relationship with her mother's practices, even after a time of rejecting it. Curanderismo turns away from Western epistemology in its rejection of Western medicine (at least for minor illnesses and the initial stages of major illness) and seeks to save itself rather than accept the salvation of modernity. In Walter Mignolo's words:

> the rhetoric of modernity is a rhetoric of salvation (by conversion yesterday, by development today), but in order to implement what the rhetoric preaches, it is necessary to marginalize or destroy whatever gets in the way of modernity. It so happens that not everyone believes in the salvation being proposed, and those who don't either react against (resistance) or engage in a critical analysis of the situation in order to move in a different direction (re-existence). (*Darker*, xxiv–xxv)

Mignolo discusses here the rejection of an idea of "salvation" via modernity's ideology of linear progress, and he argues that decolonial thinkers either resist this salvation or redefine their existence in order to avoid, or even adapt for survival, this so-called progress. In the above example, Montoya Padilla initially attempts to accept "the salvation" offered by "the book," but she eventually learns that what her mother offered her was more valuable. The women whose stories Preciado Martin preserves in this collection adapted their lives to the changes occurring in southern Arizona, but they did not give up their commitment to traditional knowledges such as curanderismo. They at least memorialize these commitments, and one hopes that somewhere a daughter or grandson or great niece has recorded their prescriptions, their recipes, and their techniques.

In another oral history, the first in the collection, Livia León Montiel tells of the lands where her family lived, worked, and loved during her lifetime. She emphasizes with great pride many times how "nothing was wasted; everything was used" (Martin, *Songs* 16) when butchering an animal or gathering crops. She also resists a consumerist value when she states, "They were not affluent times, but we never lacked for anything. We were poor, but we felt rich. We had so much that we didn't feel we needed money wealth, that dollar" (17). Here we see her rejecting a commodity-based wealth and sticking with her organic connection to the land and her family and their capacity to sustain themselves from the land itself. She describes how they experienced hardship during droughts and she remembers how they irrigated with the traditional method of acequias. Then she laments that her family's land is now a golf course and nearby there is a water park. She dryly observes: "It doesn't seem possible that there was no water for our homestead and now that park is there and they have water galore" (7). Thus she marks her family's practice of living within limited means.

The women tell us the story of the Mexican American communities of southern Arizona that underwent a complete transformation from a self-sustaining culture living off the land and in continuity with generations dating back to eighteenth-century settlements in Arizona to these same families' exile to urban centers, far from their ranches and gardens, where their carefully cultivated rural survival skills met with little appreciation. In short, their lives spanned the transformation of southern Arizona from a rural, ranching, and local economy, where people lived close to the land, to a modern and postmodern capitalist economy where the land turned into an abstracted space for the production of profit for the relatively few at the top of the new economic hierarchy. But this did not occur unperceived and their narratives show their resistance and adaptation with which they responded to this situation. In the next section, I discuss *Borderlands/La Frontera*, which also perceives a culture that has made its own adaptations and adjustments in response to modernity and coloniality.

CONCLUSION: GLORIA ANZALDÚA'S *BORDERLANDS/LA FRONTERA* (1987), THE POSTMODERN COLONIALITY OF BEING, AND THE LAND

Anzaldúa's work is understudied in regard to its environmental aspects. Integrated into her landmark work *Borderlands/La Frontera* is a retelling of border history going back to the pre-Columbian era, and throughout the text she makes it consistently clear that this one place—the Rio Grande Valley, the U.S.-Mexico borderlands—informs her theory of a new consciousness available for any kind of physical or psychic border dwellers. In the most famous and most anthologized portion of her book, the final section of the final chapter of the book's essay section that details "*La conciencia de la mestiza* / Toward a New Consciousness," she concerns herself with the experience of the land as it connects to her battered mestiza consciousness: "This land has survived possession and ill-use by five countries . . . It has survived Anglo-Mexican blood feuds, lynchings, burnings, rapes, pillage" (112). Here she considers the exploitation of the land as parallel to the oppression of its peoples, commenting first on the way the land has endured five political regimes and then using terms of human suffering to describe the land's as well as its people's exploitation. Her concern with the land in this final section continues her unease with the status of nature throughout the book. However, this subsection, titled "El retorno" (The Return) describes her experience of going back to the Rio Grande Valley after leaving it to pursue her education and career. Therefore, this section is arguably a climax of her description of the way that her environmental awareness helped her see the human crisis in this region, as well as how the land's exploitation relates to the injustices she finds in the human community.

Anzaldúa opens this section by evaluating the status of the Rio Grande's waters: "I stand at the river, watch the curling, twisting serpent, a serpent nailed to the fence where the mouth of the Rio Grande empties into the Gulf" (111). Upon her return to South Texas—a return that inspires her writings—she makes it a priority to visit the river. However, instead of an idyllic communion with its waters or its surrounding landscape, she offers the violent image of "a serpent nailed to the fence." To make more sense of this description, we must recall that in her earlier chapter, "Entering into the Serpent," she associates Earth with a snake (as noted earlier in this chapter). From this comparison we can see that she associates the image of a serpent with Earth itself—the natural environment. But she does not stop there. She also extends the metaphor to herself—she finds her bodily materiality in the act of "enter[ing] into the Serpent," and in so doing, she realizes her direct connection to the animal world and the greater natural environment.

It is no wonder why she calls South Texas a "tragic valley," because there she finds poverty and environmental crisis linked together. When she returns, she finds the serpentine river—a symbol of the region's natural environment, including humans—in a trapped position, nailed to the responsibility of life as a destructive political border in a land that had been historically and bioregionally linked and marked as interdependent by the same river. Now that interdependence results in a negative impact: "The borderlands depression that was set off by the 1982 peso devaluation in Mexico resulted in the closure of hundreds of Valley businesses . . . Because the Valley is heavily dependent on agriculture and Mexican retail trade, it has the highest unemployment rates along the entire border region; it is the Valley that has been hardest hit" (112). She laments the hardship she finds in the valley, and she makes the same connections that González made in her work some fifty years earlier: "Some of the poorest people in the US live in the Lower Rio Grande Valley, an arid and semi-arid land of irrigated farming, intense sunlight and heat, citrus groves next to chaparral and cactus" (Anzaldúa 111). She draws out the contrast between maintaining citrus groves in an arid/semi-arid bioregion where chaparral and cactus grow more naturally, and she knows this personally because her father was a farm laborer who would spend some of those dry, scorching days "scanning the sky for a rain that would end the drought" (112).

By the time Anzaldúa returned to South Texas in the 1980s, farming irrigation had depleted and polluted the Rio Grande. The promotion of South Texas as a farming region also motivated a trend to clear the land to make way for farms—some of which never materialized. This resulted in clear-cut lands that exposed topsoil to destructive erosion. South Texas newspaper reporter for the *McAllen Monitor*, Arturo Longoria, observes, "Soil erosion in deep South Texas had become so extensive in the late 1980s that the National Wildlife Federation called the Rio Grande Valley one of the most severely wind-eroded areas in the United States. The NWF research concluded that an average of fifteen tons of soil per acre was being lost annually in many parts of the region" (62).

What Anzaldúa intuits about the natural environment and observes in the Valley's impoverished population is what earlier writers anticipated. Longoria simply provides the hard facts of the destruction, and admires the lost ways of his grandfather, who "lived off the land with never a thought for want, for he abided in harmony with nature" (107). What Anzaldúa and Longoria express resonates with all the other writers in this chapter: Chicanas and Chicanos who know that our culture maintains a unique relationship with, not dominance over, nature. More radical than reclaiming lost title to lands, these writers declare that our communion with nature ranks higher than any legal document, even if we are the only ones who recognize it.

4

ACTIVE SUBJECTIVITY IN MIGRANT FARMWORKER FICTION

Rejecting Alienation from the Land

THE SUBGENRE OF CHICANA/O WRITING about migrant farmworkers merits its own chapter, both for the power of the writing and the ongoing vulnerability of these laborers to environmental injustice and serious abuse. One might even say that farmworkers do battle on the frontlines of the abstraction of land in the modern and postmodern capitalist economy discussed in the previous chapter, all while providing us with the food for our tables. A little-known short essay by Denise Chávez on her experience of rafting the Colorado River provides an introduction to the specific issue of Mexican American labor and the land. Chávez was inspired to write the piece after her curiosity led her to spend some time reflecting on why so few Mexican Americans indulge in such recreations. I follow up with an analysis of a trio of works that span the 1970s through the 1990s. They reveal how farmworkers maintain intimacy with the natural environment, despite the efforts of capital to alienate workers from the land.

DENISE CHÁVEZ: MEXICAN LABOR AND "MANAGEABLE THRILLS"

Rarely do Mexican American writers venture to narrate a wilderness adventure, and in her essay Chávez defies the convention of nature writing by emphasizing labor over leisure. In "Crossing Bitter Creek: Meditations on the Colorado River," Chávez claims: "I may be the only Mexican rafting the Colorado. Everyone else is at Disneyland or in Las Vegas" (67). This statement implies a concern for leisure, and perhaps even a touch of criticism toward Mexicans who choose to spend their vacations in inauthentic and overdeveloped settings. However, she moves on to express empathy for hardworking Mexican Americans quickly thereafter. Chávez's essay appears in the 1998 anthology, *Writing Down the River: Into the Heart of the Grand Canyon*, which collects ten essays written by women who completed the 280-mile rafting trip down the Colorado River. Other than being a well-established writer of compelling novels and drama for many years, it's not clear what led to Chávez's invitation to participate in this project. One does not expect to find Mexican American writers in such a collection, especially not during the 1990s, when the links between Mexican American experiences and environmentalism were minimally documented.

Whatever the reasons, her readers are fortunate that she participated. Her essay offers a beautiful encounter with the conventional nature writing setting, but she narrates her experience from her position as a working-class Chicana feminist. That she narrates her experience from this intersectional identity means nothing less than that, in just a few paragraphs, she maps the complex geography of what it takes to make visible the environmental relevance of Mexican American writing and culture. By the 1990s, Mexican Americans focused on social justice issues and lost sight of their unique relationship with the land. At the same time, the environmental movement did not perceive interest from Mexican Americans. In the above quote, Chávez bluntly speculates that no other Mexican American shares her experience of running the Colorado at any point along the 280-mile route. Instead, she suggests, they can be found in the orchestrated, controlled entertainment environments of Disneyland or Las Vegas. Should Mexican Americans take offense at reading this speculation? On the contrary, Chávez performs a series of necessary interventions, within the realm of environmentalism in general and nature writing in particular, on behalf of working-class Mexican Americans. Specifically, she discusses how class, environmental justice, history, and decoloniality are brought to bear on this situation.

What keeps Mexican Americans from participating in such river adventures in greater numbers is not a lack of concern for the status of the natural environment, nor do we lack sensitivity to the lessons that can be learned from facing the challenges of a rafting trip. Chávez initially considers the prohibitive cost of such a journey, but she knows that money would not be a barrier for everyone. Nonetheless, her first intervention concerns class: she recognizes the impact of income levels on everyone's access to ecotourism. She continues with her second and related intervention by offering an alternative explanation: "We are a people who want comfort, manageable thrills, self-made happiness" (68). Here Chávez shows how Mexican Americans avoid the risky randomness of adventures in nature and instead opt for predictable pleasures. The insights of environmental justice can help interpret this statement. Environmental justice, defined as "the right of all people to share equally the benefits bestowed by a healthy environment," exposes the fact that some people bear the unequal burden of the unpredictable aspects of nature in their everyday work and home settings (Adamson, Evans, and Stein 4). Why would the people of this population then choose to expose themselves to further risk during their rare moments of travel and leisure? Once seen through the lens of environmental injustice, the working-class Mexican American reluctance to raft the Colorado makes more sense.

Chávez does not stop with environmental justice. In the next sentence, she defends this orientation and offers the insights of her historically informed, decolonial perspective, which make up her third and fourth interventions. "Our blood runs too deeply down the muddy river. We were and are its slaves, its laborers, its workers, its drones. The memory of hardship is still too much to bear. Our ancestors tried to conquer this river, but they turned back too soon" (68). Here she shows how Mexican Americans, and ethnic Americans in general, cannot help but see the beauty of the river mixed with the violence of the land it runs through: "our blood runs too deeply down the muddy river." The racial hierarchies, slavery, and oppression instituted by colonization are never far from the minds of ethnic Americans in such settings. Not only does this statement bolster the environmental justice view of the previous sentence, it also recalls images of the racial impact of conquest and invokes the deep colonial history of the U.S. Southwest: "the memory of hardship is still too much to bear." By insisting on this view, Chávez makes a historical, decolonial intervention in conventional nature writing. In other words, she refuses to dwell only on the present beauty of the surroundings. Instead, she shows how Mexican Americans live with the legacy of Spanish colonization as well as the U.S. imperial takeover of half of Mexico's territory: "our ancestors . . . turned back too soon."

FIGURE 4. *Mi familia* weeding sugar beets in Michigan, circa 1940. Photo from author's personal collection.

One cannot decipher the message of Mexican American writing and the environment without considering the colonial and imperial histories of the United States. What makes this essay decolonial is that Chávez chooses not to dwell in a place of resistance. In fact, to understand the ecological lessons of Chicana/o cultural productions, one must see them through a decolonial lens. The decolonial landscape situates Latin America's locus of knowledge in its own geography, place-time, and peoples (indigenous and mestizo/hybrid). Chávez champions this approach in her short essay, and the following writers show how it works in their fiction about farmworkers in the latter half of the twentieth century.

ACTIVE SUBJECTIVITY AND THE CHICANA/O FARMWORKER NARRATIVE

When it comes down to it, the two topics of Chicana/o writing that most obviously connect to environmental issues for me are curanderismo and the migrant farmworker experience. My mother raised me with the care of homeopathic recipes that she learned during her childhood in rural Mexico, and my father often told me stories of his farmworker experiences during the Great Depression, following the same Texas-Midwest-Texas circuit that Tomás Rivera's protagonists trace in *And the Earth Did Not Devour Him* (see figure 4). I have taken up many different ways that Mexican American and Chicana/o writing connects to environmental issues in this book because I felt a need to historicize and contextualize, especially within coloniality and capital, these traditions and experiences. I addressed curanderismo in the previous chapter. In this chapter, I address the depictions of the migrant farmworker experience in three exemplary works from the second half of the twentieth century. Many works comprise what is arguably a subgenre of Chicana/o writing—farmworker writing—but here I will take up only *And the Earth Did Not Devour Him* by Tomás Rivera, "Something in the Rain" by singer-songwriter Tish Hinojosa, and *Under the Feet of Jesus* by Helena María Viramontes. *And the Earth Did Not Devour Him* is in some ways the most iconic novel in this subgenre, with its relatively early publication date (1971) and the masterful way Rivera represents the diverse community of migrant farmworkers from Texas in the 1940s. "Something in the Rain" offers a lesser-known Chicana environmental justice contribution to popular culture. *Under the Feet of Jesus* is the work of Chicana/o literature most often the subject of environmental justice ecocriticism.

In this chapter, I argue that migrant farmworker literature shows this community's rejection of capital's attempt to alienate them from the land. The migrant farmworkers in these writings assert a community with one another and with nature. Some consider the discussion of a Chicana/o relationship with the land to be a romanticized narrative that should be left behind.[1] However, if one stops to consider all the danger and difficulty that mediates the farmworker's experience of the natural environment, the fact that these narratives include appreciation of the places where they live and work testifies to the farmworkers' capacity to still see the land despite capital's attempt to render it abstract and alienating. If, when we read a farmworker's appreciation of natural beauty, we hear it as "romantic," then we deeply underestimate the farmworker's capacity for creatively engaging his or her surroundings and his or her understanding of basic ecology. We also misunderstand the farmworker's deliberate disconnection from the legacy of modernity. When the farmworker appreciates natural beauty, he or she does so from the knowledge gained from daily effort and relationship, rather than from a trite imitation of a long dead poet or philosopher. In order to better understand the expressions and narratives by and about farmworkers, I use the decolonial theories of María Lugones.

Lugones strives to name actions of subjects who purposefully disconnect from dominant systems of power and hierarchy. What we see in these narratives is the way farmworkers enact what Lugones characterizes as "active subjectivity." She defines active subjectivity as "highly attenuated" and as a practice that "does not presuppose the individual subject and it does not presuppose collective intentionality of collectivities of the same" (Lugones 6). Her theory of active subjectivity aims to describe the actions of multiply oppressed peoples in a way that does not impose expectations of direct action in order to earn the title of subject rather than passive object. In other words, she opens the interstices between subject and object, active and passive, by theorizing the subject position as an ongoing activity of negotiation rather than as a state of being. This gives the oppressed subject the space to navigate multiple oppressions and identities. She states, "I am keen on theorizing what, from the standpoint of liberalism, would look like an almost inconsequential or attenuated sense of agency" (5).

With this conception of decolonial subjectivity, Lugones offers an alternative to liberalism's expectation of action and the absolute dichotomy between active and passive. This becomes vital in reading migrant farmworker literature because this community has so little access to power and thus quite a tenuous platform from which to act. But this does not mean they do not possess the capacity to respond to their situation in a significant way. We have not had the lens through which to detect the creativity of their responses and to gauge the relationship between their internal processes of awareness and their external demands for change. This chapter will show how this active subjectivity operates in these three works not only to respond to coloniality's impact on the human community, but also to respond to how coloniality commits violence against the land itself, as portrayed in migrant farmworker writings. The farmworker's ability to see and appreciate the land offers a critique in and of itself. Why do those who enjoy the privilege of living removed from the most difficult environmental circumstances make some of the worst ecological decisions and why must they be exhorted to develop a relationship with nature by spending top dollar for wilderness tours? Why do those who endure the harshest aspects of the environment consistently express their concern for the prevention of climate change and highly value their local environments, even if we don't always find them on whitewater river rafts and remote mountain trails? These three works give insight to these questions, and show how active subjectivity works in negotiating goodlife values in desperate circumstances.

TOMÁS RIVERA: "THE SPOKEN WORD WAS THE SEED OF LOVE IN THE DARKNESS"

While *And the Earth Did Not Devour Him* (hereafter *Earth*) seems a novel of suffering, violence, and the loss of all but a mere glimmer of hope, it also shows optimism in the way Mexican American farmworkers survive with their relationship to the land intact. *Earth* offers a narratively inventive representation of migrant farmworkers' lives in the 1940s. Rivera challenges readers to walk in the workers' shoes, especially from the perspective of a young boy.

His novel depicts the stark and barren lives of migrant farmworkers traveling between Texas and the Midwest during the 1940s. The hardworking characters in this poignant story struggle to maintain a positive relationship with nature. The characters' success in this endeavor becomes a telling triumph in light of the cruel treatment from others that they continuously endure. They suffer proximity to harsh elements of nature—hot sun, dirt, barren living quarters, thorny plants, pests, hard-to-reach vines, little access to clean water—in order to provide food for the very people who refuse to pay a livable wage or provide them with just working conditions. For their part, the migrant farmworkers refuse to reject nature, taking into account both its harsh and its nurturing elements, choosing instead to appreciate that it gives them a way to provide for their needs. Still, their labor in the fields that feed the nation goes unrecognized: they live in deplorable conditions, with most families sleeping in tents, old barns, and former chicken coops. They receive low pay, work in unpredictable seasons and places, with children often missing school days, and the communities into which they arrive treat them with suspicion and contempt.

For a novel written and published in 1971, during the vibrant Chicana/o Civil Rights Movement, *Earth* curiously dwells on the haunting past of the 1940s rather than painting a vivid portrait of its present or future. Indeed, my students sometimes complain about the novel's relentless pessimism. However, this novel offers a unique opportunity to honor the difficulty of life prior to the Civil Rights Movement, while it also provides a critique of the legacy of coloniality and modernity in the form of organized religion. The novel directs the community to reject what it sees as superstitions, especially Christianity, which keeps the people powerless, and suggests they work together to assert their rights. Hope glimmers, albeit faintly, when the farmworkers refuse to allow organized religion and oppressive communities to determine their relationship with nature. Only with this wisdom does the community emerge from its powerlessness. The narrative leads to insight about how the people can reject Christianity and help themselves via an organic approach with storytelling.

Indeed, with this storytelling, they enact their active subjectivity by narrating their lives for themselves instead of mimicking the tales imported by modernity and coloniality. They piece together home, healing, and comfort from the nature that surrounds them, implicitly rejecting consumerist aspirations. In this manner they move away from exploitation—of themselves and of the land—and toward empowerment. When I first read the novel, these stories were familiar to me in voice, content, and style. I grew up listening to descriptions of the migrant farmworker life from my father, Encarnación Peña Ybarra; his niece, Manuela Chavoya; and my grandmother, Apolinaria Peña Ybarra (see figure 4). That Rivera originally wrote the novel in a colloquial Spanish unique to Texas-Mexican Americans, and that sounds like home to me, makes even clearer the idea that this novel aims to depict the community's own ways of facing their plight and maintaining an alternative narrative.[2]

By no means, however, does the novel present these messages in a didactic or simple manner. The characters in *Earth* constantly find themselves on the road looking for employment, and in a narrative sense, readers wander alongside them. By definition, migrant farmworkers follow seasonal crops around the country, settling in one place for short periods at a time and often not knowing where they will next land. In a recent study, anthropologist Seth M. Holmes documents this process, commenting that his book is an "attempt to present the unfolding narrative nature of the experience of migration . . . [and] demonstrating the everyday joys and suffering involved in migration as well as the bodily experiences of multisited fieldwork in transit" (27). Rivera lets readers experience this mobility and uncertainty by employing a unique narrative style.

The book consists of many short vignettes, some only a few lines long, others simply dialogues with no concrete indication of who speaks the lines. No central protagonist links all the stories, although readers do tend to rely on the recurring perspective of a young boy as a general orientation to the novel's events—such as a fight at school, working in the fields with other young siblings, taking his first communion, and questioning the existence of the Christian God. Readers must put together the narrative from these disparate pieces, much like the characters themselves create a life from a constant variation of places and experiences.

The evocative title *And the Earth Did Not Devour Him* makes reference to the Earth that sustains the community and upon which they survive by not quite getting devoured by their labors. The title takes on an additional meaning about midway through the novel, when the young boy challenges the devil in a vignette titled "A Silvery Night." Driven by curiosity, he visits an empty wood at midnight and summons the devil. When the devil does not appear, he at first experiences relief. But then his thoughts wander to the next logical conclusion: if the devil does not exist, then neither does God. He can't quite bring himself to say aloud that God does not exist, but he does declare to himself: "*There is no devil, there is nothing*" (106, emphasis original). A child raised in strict Christianity and in harsh living conditions fears the devil more than he sustains a relationship with God; on this moonlit night, the boy discovers that "*there is nothing*" to scare him, but there is also no more sanctified protection for him, either. The novel does not regret the loss of God. The vignette immediately following belies a deep cynicism about Christianity. It concerns a carpenter and a minister's wife who use their "charitable" project to "help" the farmworker community as a cover for their extramarital affair, when in fact they do nothing to help the migrants. Together, these two vignettes clearly condemn religion, but the novel includes a third vignette with the same title as the novel to seal the deal.

"And the Earth Did Not Devour Him" concerns a young boy whose father suffers heat stroke while picking crops. The boy observes his mother's practice of candle-lit prayer but he does not see his father improving. At this point, the boy explodes at his mother: "How come we're like this, like we're buried alive? Either the germs eat us alive or the sun burns us up . . . All the time feeding the earth and the sun, only to one day, just like that, get struck down by the sun. And there you are helpless . . . why, God doesn't care about us . . . I don't think there even is . . ." (109).

With this outburst, the boy condemns his mother's faith in God as useless, and he almost reveals to her his new knowledge of God's nonexistence. The next morning, instead of staying home to care for his father, the family's need compels the boy to venture back into the fields with his younger siblings. His nine-year-old brother soon succumbs to the heat and sun that first took their father. The weight of his young brother passed out in his arms drives the boy to an extreme: he curses God. Strangely, as with the devil, the unexpected occurs: "For a second he saw the earth opening up to devour him. Then he felt his footsteps against the earth, compact, more solid than ever. Then his anger swelled up again and he vented it by cursing God. He looked at his brother, he no longer looked sick" (111).

Instead of experiencing the wrath of God at his curse, the boy finds the situation improving immediately. The Earth does not serve God's will; it stands firm under his feet. His relation with the land is reinforced while his and the land's subservience to God is called into question. The next morning his father feels better and the boy takes joy in "the freshness of the morning" (111) and feels "for the first time . . . capable of doing and undoing anything he pleased" (112). His rejection of God inspires him to renew his commitment to his family's endeavor with a fresh spirit no longer weighed down by feeling intimidation and impotence in the shadow of the Christian deity. Importantly, the boy and the Earth become allies. He finds strength in his tread upon the Earth, and this strength helps him fight against hardship. His growing alliance with nature is foreshadowed when he first discovers the devil's and therefore God's absence. His last thoughts of the day concern these absences, and he falls "asleep gazing at the moon as it jumped through the clouds and the trees, as if it were extremely content about something" (106). The moon celebrates the boy's alliance with nature, even as he falls into his Earth-bound sleep.

Still, rejection of God and Christianity does not automatically lead to collective action against oppressive labor conditions. The boy voices his frustrations and gains a sense of power, but he does this alone, even alienating himself from his mother. The novel suggests a possible strategy for collective emancipation by showing the importance of orality in Mexican American migrant farmworker culture. Their orality results from a deeply regrettable circumstance: the farmworkers are often illiterate, having attended few days of school due to their destitute circumstances, child labor, and migratory ways. However, another side exists to this orality, and the novel attempts to represent it in many different ways. The novel includes many instances of dialogue with no speaker or setting indicated. At other times, the narrative addresses the reader in the second person, as if involved in a personal conversation. Still other instances show how tales pass from one person to another, with the speaker shifting as the tale changes in his or her particular version. This emphasis on orality showcases one of the ways this community can help itself: by narrating its own experiences.

A good example of stories passing among different people is the tale of the thwarted lover. Titled "The Night the Lights Went Out," the vignette tells the story of a love triangle from many different perspectives, allowing the characters themselves a voice in the telling, too. One voice says, "That Ramón, he loved his girlfriend a lot . . . They say it was the first time in four months that he had seen her," and later another voice claims, "They say she danced the whole night with Ramiro," while yet another pitches in, "Well, they say Juanita asked her parents for permission to leave early for the dance" (126). Each of these narrators declares from the outset that he or she heard the story repeated by others: "they say." The tale ends with Ramón sacrificing himself to his forsaken love.

One might interpret this story as a misogynistic judgment against women's whims, while another might see it as a critique of gossip within a community. Yet what stands out is the manner in which the story is told—why include so many different voices if not making a point about the power of passing around stories? Nothing explicitly faults the woman for her role; the story simply shows the tragedy of lost love in the context of hard work. The newcomer Ramiro dressed well and did not join Ramón on the migrant trail; his access to more stability and money overshadowed the hardworking Ramón. More than anything, the community circulates the story to show how an individual who tries to elevate himself alone above the rest only causes tragedy. They comment about Juanita's love interest, Ramiro: "He was nothing but a show off and he was always all duded up. They say he wore orange shoes and real long coats and always had his collar turned up" (125). The community must cultivate more collective strategies—via an oral tradition, for example—in order to improve its lot. They already have an oral tradition and now they must use it to work toward a better life together in social action. Some critics may consider this analysis that emphasizes the community-oriented values in this novel a "tired theme," but collective action remains one of our most vital strategies in the context of coloniality (Gallego 33).[3]

The most poignant tale concerning the oral tradition comes at a key point, next to last in the novel, and provides one of the few respites in the otherwise harrowing narrative. It is a five-sentence vignette that tells of Bartolo, the composer and reader of poems. He visits various communities and composes poems on the spot for whomever pays him, and he includes their names in the poems, much to their delight: "when he read them aloud it was something emotional and serious" (Rivera 147). Enjoying little direct agency in their lives, the farmworkers take joy in hearing poetry in which they play an active role. Indeed, the story ends with this insight: "the spoken word was the seed of love in the darkness" (147). Although achieving literacy for all will go a long way to help this community, they cannot wait for that day to arrive before they take action. Tell your stories now, the novel admonishes, to one another, and work together to create change. Otherwise, neither God, nor the clergy, nor anyone really, will help you in your silence. This is an example of the active subjectivity that can ultimately lead to action.

Given that humans gain environmental knowledge about a place over long periods of time, one would expect migrant farmworkers to have a superficial relationship with any of the places where they happen to live and work. However, characters in this novel display a kinship with their locales that belies their constant exposure to harsh environments and their temporary, migratory relationship to various places. Indeed, in a sharp critique of place-based ecocriticism, Janet Fiskio insightfully observes that migrant farmworkers contribute a very important element to our current understanding of agriculture by "conceptualizing place as a capacity and practice rather than as a static ideal" (320).[4] Fiskio also adds, "The landscape of agriculture is in the midst of transformation in the United States, and the most innovative actors are those from the margins of the food system, communities who are deterritorializing agricultural and urban landscapes and reterritorializing these spaces as site of collective action for food democracy" (320).

The laborers, and place-makers, in Rivera, Hinojosa, and Viramontes reject individualistic consumerism and honor hard work, indicating that their voices can be the most useful weapon against oppression. In alliance with the Earth that does not devour but strengthens, they can work together toward change. Tish Hinojosa's song reveals the ongoing difficulties that farmworkers face. She penned her lyrics in the 1990s and, tragically, little has changed in regard to the farmworkers' situation in the fields or in their living conditions.

TISH HINOJOSA: "TALKERS TALK AND DREAMERS DREAM"

In "Something in the Rain," singer-songwriter Tish Hinojosa tells the story of a little girl's death as a result of pesticide contamination from the naïve perspective of her older brother. Helena María Viramontes's *Under the Feet of Jesus* is perhaps the most well-known instance of writing about farmworker pesticide poisoning, and I will address it in the final section of this chapter. Because many scholars have written about the environmental injustices depicted in Viramontes's novel, I want to first introduce this song as a lesser-known text in this category.

Leticia "Tish" Hinojosa grew up during the 1960s on the west side of San Antonio, Texas, which is the low-income Mexican American barrio of the city. She was the youngest of thirteen siblings, and both her parents were immigrants from Mexico. Her mother aspired to be a professional singer but never had the opportunity to achieve her dream. Hinojosa first started singing and playing acoustic guitar with her family and then in local bars and coffee shops before she moved to New Mexico and then to Nashville, where she discovered her love of songwriting. Hinojosa records and writes in a variety of musical genres—folk, country, singer-songwriter, Tejano, Americana, and corrido, to name a few. She has been producing records since 1987, and so far her discography numbers sixteen. As a successful recording artist, she has toured the United States and Europe several times. Her recordings include songs in English and Spanish, and some records were produced in separate bilingual editions. Her songs cover a range of topics from the typical love stories and traveling songs to forays into academic topics, such as her song dedicated to the career of don Américo Paredes ("Con Su Pluma En Su Mano" [With His Pen in His Hand]), and a recording inspired by Octavio Paz's *Labyrinth of Solitude*, a classic text of Mexican philosophy published in 1950.

"Something in the Rain" appears on Hinojosa's 1992 Rounder Records outing *Culture Swing*, which includes an eclectic mix of Tejano sounds, country tunes, a song originally recorded by a Spanish artist, and western swing. "Something in the Rain" is a folk song, with its poetry telling the tragic story of a girl's death due to pesticide poisoning and her brother's coming into awareness. The narrative drives the structure of the song, making it more of a poem than a conventionally organized lyric written to a melody. It is the only song on the album that takes on a social justice issue. It begins with melancholy guitar strains and continues in a thoughtful and reflective tone. On one level, this song exposes this boy's ignorance and naïveté when his sister becomes ill, and then on another, it shows that he realizes what has befallen her and begins to think about how he can try to prevent other deaths like his sister's. In that sense, it can be interpreted as a rallying cry for activism against pesticide use. Yet not one line confronts the listener with an exhortation to action, and the song's quiet tone suggests a subtler message. The song gives us a window on this boy's reality. It also offers insight into the challenges a farmworker family faces and how those challenges can prevent them from reaching a decision to become activists and to resist their abusive situation. A deeper look into the song's lyrics and musical arrangement reveals this family's careful negotiation of a complex web of possible agency and certain threat to their existence. After all, they lose a daughter by the fourth stanza. The song represents different levels of faith, knowledge, and action. Once the young narrator sorts through these three realms, he emerges at a place of active subjectivity.

First, the song demonstrates the persistence of faith in this community, despite the skepticism of the civil rights era evidenced in *Earth*: "When troubles come our way / Oh yeah, I've seen my daddy pray."[5] The outright rejection of Christianity found in Rivera's novel does not survive to reappear in this song or other Chicana/o works of this era, showing that this community can still realize an awareness of activism without outright rejection of faith. Other types of belief or faith are also revealed. The boy's mother encourages him to "dream of someplace far from here" when his sister is dying and suggests he consider "the pictures in your school books / someday you can take us there." This belief in the redemption that education promises must, however, be tempered by the fact that he attends school only "when work is slow." His most emphatic faith comes at the end when he states, "I'm afraid but I believe / that we can change these hurting fields." But this other kind of faith comes only after he negotiates the layers of knowledge available to him.

The song reveals the epistemological hierarchy that keeps the boy's knowledge at the level of his heavy labor ("Mom and Dad have worked the fields / . . . / I'm just a boy but I know how") and keeps him from apprehending the harms of pesticides because he only infrequently attends school, owing to his employment ("[I] go to school when work is slow"). Because of his ignorance, he blames the rain for his sister's illness: "There must be something in the rain / I'm not sure just what that means." Every time, save the last, that Hinojosa sings "there must be something in the rain," she is accompanied by a harmonizing voice, which suggests this line contains an additional connotation, another layer of sound for another layer of meaning. Indeed, his grandmother hints to him that factors other than the rain are involved: "*Abuelita* talks of sins of man / Of dust that's in our hands." Even after she shares this with him, suggesting that their hands are dusted by harmful pesticides spread by the growers' planes, he still insists, "Those airplanes cure the plants so things can grow / Oh no, it must be something in the rain." His insistence reveals the way their socioeconomic circumstances alienate the boy from his natural environment and makes him, at least temporarily, fault the naturally occurring rain for his sister's death rather than the planes spreading harmful pesticides. A passionate guitar solo follows these poignant lines, indicating the passing of time and a shift in mood and perspective. By the next stanza he declares that he has gained a new knowledge and insight: "And me, I think I understand / About our life, about our land." Hinojosa sings these lines a cappella during a two-measure break in the accompaniment that indicates an emphasis on the message found here. What happens to create such a shift in awareness? The next two lines explain. His sister dies ("Little sister's gone away"), and his mother is largely absent ("Mama's workin' long again"). This dual trauma serves to make him aware of the more precise culprit. Although the pesticides may fall from the skies, they descend by no natural means but instead by the artificial "rain" created by "the sins of man / Whose profits rape the land." The pivotal moment of awareness is revealed at the same time that he demonstrates kinship with his local environment when he says that he understands "about our life, about our land." When he says "our," he declares no ownership over the land but instead avows a deep communion with it. The ending rings a plaintive note, with the boy declaring that "the rains are pouring down / from the growers to the town." He observes that the poison will reach not just the farmworkers but the "town" as well. He also declares that the artificial and poisonous "rain" will continue to wreak its destruction unless "we break the killing chains." But how?

The penultimate stanza, where Hinojosa again emphasizes the importance of these lines with another two-measure break in the accompaniment, gives insight into his strategy. He states that "talkers talk, and dreamers dream / I will find a place between / I'm afraid but I believe / that we can change these hurting fields." What does it mean to "find a place between" talking and dreaming? Perhaps the talkers are the complainers and pessimists, while the dreamers are the activists fighting for farmworker rights. This boy must face the reality of his situation: the good found by realizing his intimacy with the land and the bad found in his daily vulnerability to violence, illness, and death. He refuses to talk the talk of complainers and pessimists. But neither can he imagine the activism that might be expected of him. He offers a vision of systemic change ("break the killing chains") as the only hope for preventing the "rain" from infecting "the town." Until then, he declares in the last line of the song, "there's something in the rain," and Hinojosa leaves this line without a harmonizing accompaniment, suggesting that the boy now understands where the poison originates. In other words, he can see the benefits of the dreamers' dreams to regulate pesticide use or enforce worker protections, but he suggests that the "chains" of the entire system of coloniality and modernity must be broken before real change will occur. He admits to fear but also commits to change: "I'm afraid but I believe / that we can change these hurting fields / 'Cause there's something in the rain / but there's more here in our hands."

The boy's initial impression that the rain poisoned his sister slowly shifts as the verses eventually reveal a different culprit. He finally realizes that he was mistaken to hold faith in the fact that "those airplanes cure the plants so things can grow." He ends with a pledge to "break the killing chains"—to delink from faith in modernity, which only creates more problems when it claims it "can cure the plants." Hinojosa's song gives us a brief look into the life of one young boy whose experiences are emblematic of the active subject. The final line that declares "there's more here in our hands" leaves the possible action open, but suggests that the farmworkers have something to contribute to this conflict. The boy might feel overwhelmed, but he is not helpless. The young protagonist Estrella in Viramontes's *Under the Feet of Jesus* also realizes that she can create change with her own hands and help herself, her family, and her friend Alejo.

HELENA MARÍA VIRAMONTES: "*YOU* DON'T MAKE IT EASY FOR *THEM*"

Several scholars have written about *Under the Feet of Jesus* and the environmental issues it illustrates, including pesticide use and food injustice.[6] It tells the story of a thirteen-year-old migrant farmworker girl and her family's trials as they work in California during the 1990s. This novel is the Chicana/o work that ecocritics have most engaged with, and it figures importantly in other critical contexts as well, such as labor, identity, class, bilingualism, and education. Indeed, so much has been written so well about this novel's environmental implications that I hesitate to do so again here. However, *Under the Feet of Jesus* complements the active subjectivity found in Rivera's novel and Hinojosa's song, so I comment on it briefly.

In this novel, Viramontes offers a sensitive and moving portrayal of migrant farmworker life in the 1990s without holding back a serious critique of what puts this population in such dangerous and exploitative circumstances. The novel's title refers to the place where the mother, Petra, keeps her children's birth certificates, which prove that they are U.S. citizens. With this the novel establishes from the beginning how this family's citizenship gives it a certain degree of security, but that they live in fear of harassment and deportation despite the "legitimacy" they gain from their legal status and the important role they play in the U.S. food system. The exploitation, legitimized by immigration laws, in the system of capital and in the abstraction of labor that supports the U.S. food production system becomes clear from the start.

The novel shows the fear that plagues Estrella when she happens to be nearby when the lights come on for a summer evening baseball game. Instead of experiencing the lights as part of a beloved American summer ritual, Estrella is "startled when the sheets of high-powered lights beamed on the playing field like headlights of cars, blinding her" (59). She immediately thinks "the border patrol . . . and she tried to remember which side she was on and which side of the mesh wire she was safe in" (60). She then runs straight for the shack where her family is squatting for the season, and she startles her mother with her fear. With such a tenuous psychic state—when playing field lights are trauma-inducing threats—it might take a lot of convincing before Estrella decides to take action and confront the establishment to claim more rights. Estrella embodies an active subjectivity strategy at this point in the novel in the way she is aware of her position but has not yet chosen to take direct action.

The family in this novel consists of Petra, the mother; Perfecto, the mother's boyfriend; Estrella, our protagonist; and Estrella's younger siblings, two boys and twin girls. Early in the novel, they encounter Alejo, a young farmworker from Texas. Alejo works the fields only during the summers in order to save money to help his grandmother with expenses during the school year, when he gets the kind of education Estrella so desires for herself and her siblings. Estrella's friendship with Alejo makes evident the different social positions the migrant farmworkers hold. While Estrella learns to read only at an advanced age because her school attendance is so episodic as a result of her family's farm labor following the crops and the seasons, Alejo benefits from a stable education, envisions himself in college, and even tells Estrella of his dream to become a geologist.

Given his relatively more stable social position, one might expect Alejo to become Estrella's mentor in this narrative, perhaps sharing with her knowledge about labor organizing or civil rights. But the opposite occurs. Estrella becomes Alejo's mentor and protector. In chapter 4, which many critics agree is the turning point in this novel, Estrella takes the migrant farmworker genre from active subjectivity to physical and violent confrontation, all in the name of protecting Alejo's health. The conversation that ensues between Estrella and Alejo after this confrontation shows the delicate balance of maintaining dignity and respect in the face of serious injustice. The event that tragically bonds Estrella and Alejo, and that changes Estrella's life, occurs after Alejo is sprayed by pesticides in a peach orchard.

In an attempt to make more money, Alejo and his cousin steal peaches from a nearby orchard to sell in the flea market on the weekends. No one knows that the two boys are in the orchard, and a plane sprays pesticides while they are there. Alejo is drenched in a toxic shower. The boys are in California alone, without their families, so Alejo gravitates, even more than he already had been, toward Estrella's family once he becomes ill from his exposure to the pesticides. Estrella's mother, Petra, uses herbal remedies to try to help Alejo, but her homeopathies prove no match for his illness brought about by industrial-strength chemicals. As Alejo becomes increasingly ill, Estrella grows closer to him emotionally and plans a way to help him. Perfecto has been asking Estrella to help him tear down a local barn, but she has stubbornly refused. Once she realizes she needs Perfecto's help to get Alejo to a medical facility, she barters with Perfecto. She will help him with the barn if he helps her drive Alejo to where he will need to go. Perfecto agrees.

The family makes it to a medical clinic, but only a nurse remains at the end of the day. Once the nurse perfunctorily examines Alejo, she informs Estrella that he's very sick and they need to take him to the hospital, which is twenty miles from the clinic. The nurse then proceeds to charge the family ten dollars, discounted from the usual fifteen because she knows "times are hard these days" (144). Perfecto only has $9.07, but before he hands over the money, he and Estrella attempt to barter handyman services with the nurse. She pointedly declines this informal method of payment, agrees to take the only money left to the whole family, all the while she considers herself to be doing them a favor: "Not to worry . . . Listen, a few pennies short don't mean much" (145). Even after handing over the money, Estrella and Perfecto still try to barter services and hope to regain their money because the car's gas gauge is on empty and they need gasoline to fuel their trip to the hospital and back home to this season's squatter's shack. The fee for no apparent service "seemed to [Estrella] unfair" (147).

As the resigned family begins to leave the clinic, Estrella marches to the car with purpose, even though "she didn't know what she was about to do, but had to do something to get the money for the gas for the hospital for Alejo" (148).

What she ends up doing is retrieving a crowbar from the car and threatening the nurse with it, even smashing it across the nurse's desk and destroying photos and knickknacks. Estrella demands the nurse return their money, and when the nurse hands her the lock box, Estrella carefully counts out and takes only the $9.07 that belong to her family. She does not once consider taking more than her due, resisting any temptation toward greed.

Other than her courage and her dignified defiance of greed, two other aspects of this confrontation stand out in the context of active subjectivity and goodlife values. One aspect concerns the fact that Estrella reaches her decision to move from active subjectivity to a specific action and even confrontation after she considers her position in relation to the nurse. In the microcosm of the clinic, the family owed the nurse for her services rendered. But Estrella expands the focus of their relations and realizes that the family does just as much, if not more, to help the nurse survive because their field work puts food on her table—their hard labor fuels her life: "Estrella had figured it out: the nurse owed *them* as much as they owed her" (148). It is not through acquisitive ambition or consumerist desire that Estrella takes back the clinic fee, but she reclaims the money to rectify the actual order of labor exchange, as she sees it. This action remains loyal to goodlife values in its defiance of greed and its appraisal of true labor, rather than labor's abstraction under the system of capital accumulation.

The other aspect of this confrontation concerns Estrella's sensation of doubleness: "She felt like two Estrellas. One was a silent phantom who obediently [behaved] as the mother had requested, while the other held the crowbar and the money" (150). This doubleness narrates the girl's transition from active subjectivity to physical action that produces a material result. Her awareness of this doubleness (one can't help but think of W. E. B. DuBois's double consciousness) puts her in the position to mentor Alejo about how to survive. She lets him know that some situations require one to fight. After Estrella returns to the car with the money, she sits with Alejo as Perfecto drives them to the hospital. Alejo asks her if she hurt the nurse, and this angers Estrella: "Does it matter now?" she demands of Alejo (151). As she tries to understand for herself what motivated her actions, her internal dialogue admits that if you restrict yourself to talking, "they ignore you," but "you pick up a crowbar and break the pictures of their children, and they listen real fast" (151). To Alejo, she simply explains, "They make you that way," when he expresses concern for her violence (151). He suggests that she did it all for nothing, that he's "not worth it" and that by giving in to violence Estrella "make[s] it easy for them" (152), by which he presumably means that she should not reinforce the establishment's perception of her as a savage not worth just treatment. Critic Paula Moya interprets this scene to mean that Alejo, given his social position, buys into conventional rules of behavior that favor the establishment. Moya observes that "Alejo holds fast to a set of values that considers armed resistance to the capitalist economic system as the only real 'criminal' behavior" (Moya, "Reading" 211). This would make Alejo an antagonist against Estrella. Rather than see him in this light, I interpret his conversation with Estrella as an opportunity for her to express what pushes her toward physical action and material results.

Alejo's criticism might also suggest that the novel wants Estrella to step back from violence and take a more peaceful route. After all, Cesar Chavez went on a hunger strike to take a nonviolent stand during the grape pickers' strike. However, Estrella wins this dispute, with Alejo as well as with the nurse, when she tells Alejo, "*You* don't make it easy for *them*" (Viramontes 153). Even after he insists that "they want us to act like that" and in so doing warns Estrella to protect her dignity, Estrella whispers back, "Can't you see they want to take your heart?" (153). This is the last straw for Estrella, what pushes her to take back what is theirs—$9.07—in order to protect her heart, her spirit, her self-respect, and, importantly, to save Alejo. She refuses to perform the role given her by conventional standards of behavior that refuse to acknowledge the contributions she makes to everyone's survival by means of her and her family's physical labor.

At the novel's conclusion, when Estrella climbs to the roof of the barn that she will help Perfecto demolish the next day, she once again experiences doubleness. Upon deciding to use a chain to hoist herself up, she invokes her double: "Okay, she said to her other self" (172). Which other self is she addressing here? It would seem she invokes the action-oriented Estrella when she wants to take the daring climb. She learns from her experience at the clinic, and her eventual success with getting Alejo to the hospital, that despite her limited circumstances, she can create change for herself and her loved ones. She does not aspire to dramatic or large-scale social movements, yet. Neither does she greedily look for an exploitative social position. What she does is subtle but significant: she makes visible the internal shifts that move one from active subjectivity toward social justice action. Her communion with the night sky testifies to her ongoing relation with the natural environment: "Over the eucalyptus and behind the moon, the stars like silver pomegranates glimmered before an infinity of darkness. No wonder the angels had picked a place like this to exist" (175). Of course, she is a "star," too, as her name translates to in English, and she has figured out how to "glimmer before an infinity of darkness." She might find herself in an overwhelming situation, but she feels confident she can encounter the challenges she will face.

Long a topic for Chicana/o writings, the migrant farmworker emerges as a more complex figure in these three works. The migrant farmworker is not a powerless being without political or social perception. Neither is the farmworker always a politicized activist. Active subjectivity offers another way of seeing the farmworker who is not calling comrades to arms right away. This decolonial strategy hones the awareness of multiply oppressed peoples and reveals venues of disengagement with the status quo that might also one day lead to physical action with material results. The farmworkers who populate Cherríe Moraga's play in chapter 5 offer an example of the subject in action after harsh realities awaken an entire community.

5

ECOLOGY AND CHICANA/O CULTURAL NATIONALISM

Humility Before Death in Cherríe Moraga's Millennial Writings

GOODLIFE WRITING challenges conventional Chicana/o cultural politics as much as it undermines conventional ecocritical practices. One of the primary means that goodlife writing challenges Chicana/o cultural politics is through the deconstruction of cultural nationalism, which has been a radical way for Chicana/o cultural politics to reject colonial values. Oftentimes, cultural nationalists lay claim to colonized lands in the name of a people's sovereignty. In the previous chapters, I established how Chicana/o identity delinks from coloniality by asserting a relation to the land. In this chapter, I will look at the ways that Chicana/o cultural nationalism lays claim to land in an attempt to deterritorialize the colonizer. But to what extent has cultural nationalism concerned itself with the dignity and respect of the land as much as with the sovereignty of the peoples who inhabit it?[1] Goodlife writing by Cherríe Moraga challenges Chicana/o cultural nationalism to forge communities inclusive of the land and in this way accounts for ecological concerns. Unlike earlier goodlife writings, Moraga's work dates from an era when ecological concepts circulate more dominantly in American letters. As a result, hers is among the Chicana/o writing that most explicitly addresses environmental issues. Moraga remains consistent with the decoloniality of goodlife writing by showing how the Chicana/o relation with the natural environment has been manipulated by centuries of exposure to the values of dominance and possessiveness imposed by modernity.

By bringing ecology into conversation with cultural nationalism, Moraga works toward establishing a more joyful community in what I call her "ecological vision trilogy": *The Last Generation* (1993), *Heroes and Saints* (1994), and *The Hungry Woman: A Mexican Medea* (2001). I call this an ecological vision trilogy because these are the three works in which she most specifically connects the survival and nurturance of the Chicana/o community to a healthy relationship with the natural environment. Emblematic of the Chicana feminist strategy to transcend the colonial concept of possession, all three works emphasize care rather than domination of the land and the peoples who inhabit it. Counterintuitively, death becomes the dominant symbol for this care. Moraga communicates care by emphasizing a respect for, rather than fear of, death.

These two approaches—care for land and peoples together and respect for death—work toward reorienting Chicana/o cultural nationalism by displacing domination and replacing it with reciprocal relations that provide a place for women and men, young and old, queer and straight, human and nonhuman alike. Care for the land demands humility before death, one of the most vital stages of ecological processes. The ultimate manifestation of transcending possession is letting go of life itself. This ecological vision trilogy concerns itself with the role that death plays in keeping the community together. After all, Chicana/o goodlife writing critiques dominant values, including modernity's fetishization of life. Rather than protesting death, these three works leverage modernity's own fear of death against it in order to help Chicanas/os learn strategies for building a better community for the future.

In her collection titled *The Last Generation*, Moraga delivers some of her most impassioned pleas on behalf of a respectful and dignified relationship between humans and the environment. In this work, she muses about the chances that the revolutionary values she learned will survive into future generations. Her poetry links the colonized body to the contaminated land, thus linking their fates in a way that also indicts modernity. Modernity's humans dominate the land, whereas Moraga's decolonized peoples dwell with the land. *Heroes and Saints* is arguably Moraga's best-known play. It reveals the injuries, literal and figurative, that toxic environments inflict on a small agricultural community in the central San Joaquin Valley of California. Moreover, she shows her people's unique capacity for maintaining a love for the land upon which they labor, despite the alienation from the land that agribusiness and local government inflict on this working-class community. In this work, she shows the difficulties that the community faces in coming together to resist oppression, but they unite behind their use of death to intimidate the powers that dominate them. *The Hungry Woman* reinvents key myths from both Western and indigenous traditions that warn of possible threats to forging a joyful community for the twenty-first century. She invokes the past in order to instruct the present, and in so doing she educates the contemporary Chicana/o generation about the errors committed in the early movement, especially in regard to the ultimate price the community pays when it deploys colonial-style nationalism.

These three works challenge the patriarchy and homophobia of Chicana/o cultural nationalism, arguing that such hierarchy and discrimination reinforce racial and environmental exploitation. Moraga effectively relates the most current ecological challenges to deep histories; she connects the global climate with the most private of interpersonal relations. Her works shed light on the injustice and alienation global-scaled commerce wreaks by revealing her innermost thoughts in achingly personal essays and in plays that put readers in the middle of the most intimate interactions between family members, friends, and lovers. How can these introspective works dynamically reflect hundreds of years of political, physical, and environmental exploitation? These connections become clear once we consider her work in continuity with a tradition of goodlife writing—a writing that strives for community relations that embrace dignity and respect for one another as well as for the natural environment. Moraga joins her ongoing interrogations about how queerness challenges and enhances the boundaries of the Chicana/o community with her ambition to care for the natural environment. In her collection of works, Moraga has created a way of engaging environmentalism in which sexuality, gender, race, and class are as relevant as rivers, glaciers, trees, and animals.

Surviving and thriving as a Chicana lesbian poet, playwright, and essayist, Moraga's work often narrates her pain and isolation, yet she unabashedly claims strength and courage from her life experiences. Her powerfully wrought stories about herself, her family, and her lovers line her path toward radical politics. And it is a radical politics with a critical edge. As coeditor with Gloria Anzaldúa of the anthology *This Bridge Called My Back*, she led the way in demanding that mainstream feminists listen to the voices of women of color. Her first collection of prose and poetry, *Loving in the War Years*, challenged the Chicana/o community to confront sexism, homophobia, and heteronormativity.

In the same spirit of intracultural critique, she takes up environmental issues, with some of her most striking work especially condemning the parallel exploitation of the land and the oppressions endured by the queer, female, and dark-skinned body. She combines the settings of ongoing coloniality with the sometimes shocking and always revealing negotiations of community, family, and workplace by showing the ways that Chicanas/os reject practices that reproduce oppression, even if sometimes she also has to reveal instances when Chicanas/os fail to resist exploitative practices. Moraga exemplifies the highest standards of goodlife writing by consistently seeking out relations that can create not only inclusivity but also a higher order of joy: true communion. Her own way of expressing her goals as a writer reveals such standards. She describes herself as a writer who wants "stories that bring us to our knees, return us to God, humble us into full recognition of our dependence on one another and a flourishing planet" ("Indígena as Scribe" 94). Moraga's ambition reveals a writer especially concerned with community, and she includes the natural environment as part of the community she envisions as flourishing through interdependence.

All three of these works engage with transnational dynamics and the global currents of neoliberal politics and capital that put Chicanas/os and other peoples of color in the Global South in the position of paying the world's ecological debt. Rather than dwelling on a disappointing past or present, however, Moraga's work consistently envisions the possibility for the community to achieve peace and nurturance—including peace and nurturance for queer Chicanas/os and the natural environment. Such an optimistic ambition might seem unrealistic or even impossibly naïve. However, I would like to view her writings through the lens of the "queer utopian impulse" that José Esteban Muñoz proposes in *Cruising Utopia: The Then and There of Queer Futurity*. In this book, Muñoz offers an alternative to the "academic climate . . . [that] is dominated by a dismissal of political idealism" and instead argues that "queerness is primarily about futurity and hope" (10). Muñoz theorizes that queer cultural production "is essentially about the rejection of a here and now and an insistence on potentiality or concrete possibility for another world" (1). Moraga brings her queer utopian impulse to her multiple concerns: for a culture inclusive of queers, for a culture inclusive of people of color, for a culture inclusive of women, for a culture that integrates a respect and cooperation for the natural environment. If there's a place for the utopianism of queer futurity, it's in the contemplation of environmental challenges—a world that is too often full of grim predictions.

In these days of growing struggles to forge ethical relationships in the context of both the local and the global, Moraga's consistent engagement with the theme of nationalism becomes particularly relevant. While national identity compels an ethical relationship to the local, how does it account for one's relation to the communities outside the perceived nation? Because climate change is a global phenomenon, how relevant do national boundaries remain? Indeed, debates about nationalism have been a recurring theme for Chicana/o writing ever since the 1960s. Moraga makes sense of this nationalism in relationship with environmental issues by reinventing the Chicana/o relationship to territory. What becomes most important about land for Moraga is not holding title to or possession of it but the enactment of a dignified and respectful relationship with the natural environment that sustains human survival and cultivates joy in the community. Even as she brings attention to some of the most horrific impacts of the exploitation of nature—mutilated bodies and stunted lives—her guiding ethic embraces the goodlife.

Hers is not the only contemporary Chicana/o writing that directly engages with the environment, but it is among the most challenging for readers, and in particular for her Chicana/o and environmental audiences. She makes it clear that she intentionally writes to challenge her audiences and to avoid the phenomenon of "tourist literature, written for audiences who are strangers to the cultural and political geography of our symbols, images and history" (Moraga, *Loving* 148). She writes for an audience already familiar with Chicana/o literary themes of dispossession and social inequality as well as the sexism and homophobia within the Chicano movement. She writes to challenge the boundaries of conventional environmentalism. Moraga's heartfelt work and incisive political analyses lead the way in reinventing Chicana/o nationalism and the relationship to the Earth for a revitalized Chicana/o community into the twenty-first century. Interestingly, her representation of this transformation is death.

CHERRÍE MORAGA'S ECOLOGICAL VISION: *THE LAST GENERATION*

In *The Last Generation*, a collection of essays and poetry, Moraga conjures an ecological vision of the homeland that she has longed for, a territory to which she can finally belong, and a place she can inhabit, attuned to its environmental limits. Foregrounding the way her sexuality, gender, and Chicana identity have informed her understanding of the relationship between humans and nature, Moraga produces an expanded definition of nature, or "land":

> Land remains the common ground for all radical action. But land is more than the rocks and trees, the animal and plant life that make up the territory of Aztlán or Navajo Nation or Maya Mesoamerica. For immigrant and native alike, land is also the factories where we work, the water our children drink, and the housing project where we live. For women, lesbians, and gay men, land is that physical mass called our bodies. (173)

She uses the word "land" to describe what might otherwise be called "nature" or "environment" and integrates it with the body itself. In this move, she signals her concern with two of the most significant shifts in the relationship between the peoples and environments of North America: European colonization and American imperialism. One of the most paradigmatically challenging changes occurred as European forces began the takeover of lands inhabited by American Indians; she mentions the Aztec, Navajo, and Maya. Her use of the word "land" also calls attention to the takeover of Mexican territories by the United States—an appropriation that suddenly transformed a Mexican into a Mexican American, but not necessarily into a citizen with full legal rights to the ownership of his or her rancho. The peoples subordinated by colonization and imperialism have consistently shared twin fates with the lands they inhabit. By referring to the broadly defined "rocks and trees" and "animal and plant life" as "land," Moraga infuses nature with the political history of possession and dispossession. By referring specifically to the bodies of American Indian and Mexican women, lesbians, and gay men as also "land," she connects the environmental sense of the term with any contested biological site historically subjected to violence, external control, colonization, and exploitation. This politicized context informs the way she considers land to be a reason for radical action. The negotiations over borders for respective lands are often matters that generate conflict, even wars. But Moraga also notes that land is the physical location for those conflicts; land materially suffers the violence of exploitation, even as it is an extension of ourselves. And land suffers these oppressions not only during times of war, but every day.

Who would think of land as a factory, a glass of water, or even one's own body? Someone who connects land to the quotidian notions of job, home, and health is also someone who has likely been alienated from such basic elements of life—whether because of gender, class, ethnicity, queer desire, or any other marginalizing identifier found in the very body she or he also comes to consider as "land." Moraga explains, "As a Chicana lesbian, I know that the struggle I share with all Chicanos and Indigenous peoples is truly one of sovereignty, the sovereign right to wholly inhabit oneself (cuerpo y alma) and one's territory (pan y tierra)" (173–74). For her, "land" signifies not the area associated with a nation-state, but anything from which an individual can become alienated or dispossessed on the mere basis of one's gendered or ethnic appearance to the world, or even the self one has chosen to fashion. She puts physical territory in the same context as elements that are vital to everyday life, such as job, home, and health. Indeed, she puts the right to defining one's relationship to nature in the hands of the individual rather than the state—a move that de-emphasizes the significance of the nation-state as it enables an individual's interest and care for the natural environment. One has the same right to one's body and soul as to bread and soil (*cuerpo y alma* and *pan y tierra*). Moraga's work shows the extent to which Chicana/o literature's meditation on the loss of land resonates with the losses of dignity and respect that trace back to colonization, imperialism, and the violation of civil rights that continue today. Moraga objects to the alienation of rights to land, dignity, and respect, especially in regard to gender, sexuality, and ethnicity. In short, be your queer self and lose your job; be your female self and get paid less for your work; be your brown self and get a low-paying and physical labor–intensive job. Be your brown self and drink contaminated water, and work and live with insecticide poisons; be your female self, drink contaminated water, and get breast cancer or have a miscarriage. Be your brown queer female self and lose your home, your family, your security. One's very sense of "sovereign right" stems from the daily knowledge of how difficult it becomes to "wholly inhabit oneself (cuerpo y alma)" in the continuing alienation of humans from one another and from the natural environment.

Yet, Moraga does not simply lament the alienation of the right to the body and the land. She recuperates the sense of an alienated right by recognizing the connections it forges. Though the linking of oppressed bodies with the natural environment is the product of a parallel exploitation, Moraga's poetry testifies to an intuitive yearning for kinship that such an alignment awakens; the colonized body never really lost its connection to the natural environment. Moraga's poetry also evokes her sense of loss and her longing for sovereignty as they exist alongside her demand for a land and a community to which she feels she belongs. Especially moving is her poem "War Cry," collected in *The Last Generation* (1993):

War Cry
lo que quiero es
tierra
si no tierra, pueblo
si no pueblo, amante
si no amante, niño
soledad
tranquilidad
muerte
tierra. (42)

[War Cry
what i want/love is
territory/land
if not territory/land, community
if not community, lover
if not lover, child
solitude
tranquility
death
territory/land.]
(my translation)

"Lo que quiero es tierra," can be translated as both "What I want is territory" and "What I love is land." This line's double meaning thus demands social justice at the same time that it expresses environmental kinship. With its reference to territory, this poem sounds a political plea for the land that was taken from the indigenous peoples of North America, and for the land that was subsequently taken in 1848 and 1853 with the Treaty of Guadalupe Hidalgo and the Gadsden Purchase: "what I want is territory." But its stark lines also give voice to the complex emotions that are internal to this claim for justice: "what I love is land." The desire for kinship—the longing to belong—drives this poem through a series of intimacies. It first articulates desire for the land. When the land is denied, the desire remains, though redirected, "if not land, community." The speaker longs for different kinds of relationships—with a community, with a lover, with a child—in succession, with one following when the other is withheld. The desire for kinship remains as consistent as the land—the land that eventually becomes the only companion along the inevitable trek toward "soledad / tranquilidad / muerte" (solitude / tranquility / death). The succession of relationships demonstrates not only the unwavering desire for connection, but also a firm commitment to resistance. Even when a people is denied territory, it will forge a community: "si no tierra, pueblo" (if not territory, community). Once the community is dispersed, one might still maintain a family of two: "si no pueblo, amante" (if not community, lover). And when the lover is also denied, one can at least raise the child with a memory of the community and the lover: "si no amante, niño" (if not lover, child). And always the land is the companion along this journey of resistance. Kinship with the land is never separated from the significance of a community, a lover, and a child. Indeed, tierra completes the circle—for upon death, we are interred and eventually become one with the Earth. Tierra frames and defines the terms of the poem; tierra remains a consistent comfort that initiates the longing and survives to the end.

In "Ni for El Salvador," a work of mixed genre—both poem and journal entry—also collected in *The Last Generation*, Moraga again contemplates the four elements (land, community, lover, child):

> Ni for El Salvador
> *I am a woman nearing forty without children.*
> *I am an artist nearing forty without community.*
> *I am a lesbian nearing forty without partner.*
> *I am a Chicana nearing forty without country.*
>
> *And if it were safe, I'd spread open my thighs*
> *and let the whole world in*
> *and birth and birth and birth life.*
> *The dissolution of self, the dissolution of borders.*
>
> *But it is not safe.*
> *Ni for me.*
> *Ni for El Salvador.* (41)

She goes on to state, in the paragraph immediately below these lines, her goals as an artist. Her ultimate vision is to reach the point at which it is "safe" as described in the lines—safe enough to "spread open my thighs" and to dissolve borders of all sorts. The act of spreading "open [her] thighs" signals all at once an act of trust, love, motherhood, sexuality, eroticism. It is a deeply intimate act. This comprehensive, and inspirational, ambition creates such a vision that she offers its analogue, as a cultural critic. She wants to be an artist "who can create a theatre, a poetry, a song that dares to expose that very human weakness where we betray ourselves, our loved ones, even our own revolution" (41). Both statements reveal a sensitive and deeply personal commitment to working toward a culture that will consistently examine its values and practices to eliminate exploitation of peoples and lands, even when that means possibly betraying the cohesion of a movement. The nation-state mentioned in the title of the poem, El Salvador, was just bringing an end to its twelve-year civil war when *The Last Generation* was published in 1993. The inclusion of this wartime conflict alongside such intimate musings once again establishes the links between the personal and the political in Moraga's work. The nation-state structure, the human-made political boundaries, keep her and El Salvador from safely embracing the world and birthing life.

Indeed, in "Ni for El Salvador" Moraga, running the risk of denouncing her U.S.-based audience, dissolves borders by broadening the scale of environmental justice from the domestic to a global level when she contemplates a Central American nation, especially a Central American nation that has been in turmoil for so long, in part due to U.S. foreign policy and interventions. This is a courageous act, since she exposes her own complicity in the environmental injustice that sustains First-World consumption. Yet, she still claims that "[no] sustainable development is possible in the Americas if the United States continues to demand hamburgers, Chrysler automobiles, and refrigerators from hungry, barefoot, and energy-starved nations" (172). She suggests that Chicanas and Chicanos, who (after all) live in the United States, realize the role our nation plays in the global sociopolitical arena. Indeed, "Global capitalism . . . has been dependent on the free labor of domestically confined women and of men and women who were slaves and indentured servants—to those outside their own families and usually their own races—but the global economic framework has also depended on much of that labor to be unacknowledged and economically unawarded" (Platt 74).

The U.S. and other First-World nations' rates of consumption disregard the limits of nature in favor of convenience and comfort, with these privileges often occurring at the expense of peoples and environments in other nations, particularly nations located in the Global South. The concept of the "ecological footprint" quantifiably describes this inequity. An area's ecological footprint is calculated using several interdependent factors, including consumption rates, biological carrying capacity, and population. These factors combine to estimate the area of natural resources that a nation or a region requires to maintain its current rates of consumption for its population. Some areas consume more than their local resources can provide, and no other nation is a larger culprit on this front than the United States—a nation whose ecological footprint alone exceeds that of the whole of Latin America by at least six times. Thus, a revolution in the material demands of the United States and other overprivileged nations must occur before anyone, North or South, can enjoy a genuinely sustainable way of life. For Moraga, this sustainability includes ethical human relations on a global scale, along with ecologically sensitive uses of nature—a connection nurtured by developing a culture that intertwines the human and the natural. These changes would make a much bigger impact on decreasing climate change than would the countless campaigns to control population in the Global South. Taking such First-World charity initiatives into account, Moraga's desire to "birth and birth life" helps uncover the hypocrisy of an overconsuming nation's supposed expertise on how to help save the planet (Moraga, *Last Generation* 41).

Moraga brings attention to the way globally scaled commerce abstracts the misery that certain populations endure in service of the overconsuming First World. She responds to such exploitations by demanding a Chicano homeland, often referred to as "Aztlán" by advocates of the Chicano Nation, as a refuge from hypocrisy and exploitation. In Moraga's own words: "A term Náhuatl in root, Aztlán was that historical/mythical land where one set of Indian forbears, the Aztecs, were said to have resided 1,000 years ago. Located in the U.S. Southwest, Aztlán fueled a nationalist struggle twenty years ago, which encompassed much of the pueblo Chicano from Chicago to the borders of Chihuahua" (*Last Generation* 151). Geographically, Aztlán would encompass the lands in the U.S. Southwest once owned by Mexico but acquired by the United States through various means—events described comprehensively by Moraga as "the theft of what was once our México and before that and still Tierra Tarahumara,

Yaqui, Seri, Pima, O'odham" (*Loving* 180). Bioregionally, Aztlán would include the Chihuahuan and Sonoran deserts as well as the fertile Rio Grande Valley, the peaks of the Sierra Nevada, the lower Gulf of Mexico's wetlands, and the beaches along the Pacific. These claims for a Chicano homeland can be interpreted as a cry for a separate nation that would seem to conflict with her pursuit of social justice, especially for queer Chicanas who have hardly figured into the Chicano vision of Aztlán. To use Alicia Gaspar de Alba's paraphrase of a familiar refrain from *Animal Farm*, in the beginnings of the Chicano struggle for civil rights in the 1960s and 1970s, "some members of the Movement were more equal than others" (127). There was no room for queers in this formula. In addition to sexuality, Yvonne Yarbro-Bejarano notes that Moraga's "emphasis on indigenous ecological concerns . . . within Chicano nationalist discourse allows the textual conflation of Indian land and female brown bodies, a move that creates space for women within the ideal national subject" (108–9). However, even though Moraga attempts to make a place for women in Aztlán, history still testifies to the oppression that women and queers suffer under any such nationalist regime.

Yet her demands persist, and what at first seems a contradiction actually demonstrates the way Moraga longs for a sovereign connection to any and all land and a joyful community more than she claims to establish a new nation. Indeed, she revises the conventional understanding of "nation" to accommodate her vision of a homeland that would include justice for its peoples as well as its lands. Moraga's "project . . . of recuperating or revisiting nationalism from a queer perspective expands the notion of social or revolutionary change to include the politics of sexuality, a dimension of liberatory politics often lost in the search for nation" (Yarbro-Bejarano 108). Hardt and Negri also testify to the progressive revolutionary potential for what they call "subaltern nationalism"—nationalism deployed by the colonized within an existing nation:

> In some respects, in fact, one might even say that the function of the concept of nation is inverted when deployed among subordinated rather than dominant groups. Stated most boldly, it appears that *whereas the concept of nation promotes stasis and restoration in the hands of the dominant, it is a weapon for change and revolution in the hands of the subordinated.* (106, emphasis original)

Moraga's revision of Chicano nationalism is a case in point. The conventional understanding of national identity relies on "a continuity of blood relations, a spatial continuity of territory, and linguistic commonality" (Hardt and Negri 95). Moraga alters all three. She shifts the terms of kinship from blood (descent) relations to not-necessarily-blood (consent) relations in a move that validates queer families and others who challenge traditional ties: "Coming to terms with [my lesbianism] meant the radical re-structuring of everything I thought I held sacred. It meant acting on my woman-centered desire and against anything that stood in its way, including my Church, my family, and my 'country.' . . . [A]ct I did, because not acting would have meant my death by despair" (Moraga, *Last Generation* 146).

She often encourages the creative invention of identity and relationships, not so much in the postmodern sense of "play" as much as a means for survival in a social context that denies and suppresses the existence of certain differences—especially queerness and multiethnicity—in pursuit of a homogenous national identity. Her writings also work against any kind of linguistic purity in the way that her work blends at least two languages at a time (usually Spanish and English), and she repeatedly displays the novel ideas resulting from linguistic juxtapositions.

Most relevant to her ecological vision, Moraga challenges the aspect of national identity that pertains to the "spatial continuity of territory." She recognizes the "dangers of nationalism as a strategy for political change" and considers ways that it might undermine her effort to bring about environmental justice (Moraga, *Last Generation* 149). For example, concerning the physical territory of her proposed nation, she considers the changing political boundaries over the years and wonders: "Is our land the México of today or the México of a century and a half ago, covering thousands of miles of what is now the Southwestern United States?" (152–53). Superficially, this line presents the conundrum of temporal scale—the question arises as to what time period this revolution speaks against. But this also offers a glimpse at the intriguing contradictions in Moraga's work regarding physical territory. She begins to question the validity of engaging with nation on the terms of the colonizers: "Aztlán at times seems more *meta*physical than physical territory" (153). Conventionally, a citizen belongs to a nation and identifies with a politically and physically defined territory. In Moraga's projections, a citizen and a homeland belong to one another and maintain responsibilities to one another. Or, as Mary Pat Brady puts it in her outstanding study on the dynamics of space in Chicana literature, "Moraga actively suggests an anticartography—one that does not conceive of space as a thing to be possessed or a set of rationalized relations to be mapped. Moraga offers a different concept of spatiality, in which land and bodies blend in both metaphysical and real senses, in which perception and living cannot be distinguished so easily" (139).

The citizens of Moraga's imaginary homeland, Aztlán, refuse the appropriative attitude that nation historically engenders, and instead they enact a reciprocal relationship with nature in a way that sustains a broadly defined group: "Simply, we must give back to the earth what we take from it. We must submit to a higher 'natural' authority, as we invent new ways of making culture, making tribe, to survive and flourish as members of the world community in the next millennium" (Moraga, *Last Generation* 174). This conclusion remains consistent with the values found in Mexican American and Chicana/o goodlife writings from the late nineteenth century until today.

Of course, Moraga's homeland includes a conventional notion of land as nature—the actual soil and grass and hills and streams. But her "land" also includes one's body and the relations that go with it—social and environmental. And she recognizes the activists that already work within this goodlife paradigm:

> there are examples of the Mothers of East Los Angeles and the women of Kettleman City who have organized against the toxic contamination proposed for their communities. In the process, the Mexicana becomes a Chicana . . . ; that is, she becomes a citizen of this country, not by virtue of green card, but by virtue of the collective voice she assumes in staking her claim to this land and its resources. (156)

Moraga remains committed to emphasizing a necessary relationship with the actual, physical land. And she emphasizes a grassroots definition of citizenship, not as nation-state recognition, but as the responsibility that humans and nature maintain with one another as a collective community.

HEROES AND SAINTS: THE COST OF LAND AS POSSESSION

Imagining the ideal community in *The Last Generation*, Moraga also keeps things real by depicting the hardships Raza endure in the everyday circumstances of hard work and few resources. Her play *Heroes and Saints* (1992) shows how a community becomes a collective, in action against the powers that exploit its members and the lands they inhabit and love. At once inspiring and sorrowful, Moraga's play shows that collective action sometimes comes at the highest price, especially when undertaking resistance against long-standing exploitation. What makes this work so vital is the way it explores the challenges from within that can keep the community from throwing off its external constraints. In this play, as in reality, the Chicana/o farmworker community does not exist in a vacuum, and it must constantly mediate the impact on their own culture of hierarchical values emanating from the outside and the lure of assimilation. At the same time, this play shows how the community, once it coheres, can fight back by defying what the oppressors fear the most: illness and death.

Heroes and Saints is a fundamentally disturbing play; it forces its readers and audience members to stare into the face of physical vulnerability and mortality. This play disrupts the tidy comforts of segregated labor. While farmworkers' bodies are largely invisible to the public, which benefits both financially and physically from their toil in the fields, Moraga puts a mutilated manifestation of this injustice front and center on the stage. The central character, Cerezita, is a young woman who suffers from such severe birth defects that the audience might even find the character difficult to accept. She was born as a human head without a body. The playwright asks the audience and readers to suspend disbelief as we see how Cerezita lives as only a human head. Her condition results from the toxins to which her mother was exposed when she was pregnant and in the fields as a farmworker, a situation that Moraga bases on the all-too-familiar reality in places like McFarland, California. The play, set in the fictional town of McLaughlin, California, in the agricultural San Joaquin Valley in 1988, mirrors the reality of "when in a ten-year period from 1978 to 1988, a highly disproportionate number of children were diagnosed with cancer and were born with birth defects" in McFarland (Moraga, *Heroes* 89).

These toxic conditions in the fields still exist today and a growing number of researchers document the causal link between pesticide use and birth defects in the children of farmworkers. However, these studies face formidable opposition from the chemical companies who sell these pesticides. Their lobbyists work to defend the profits of companies whose products have not been fully vetted for their impact on the human body. Doubtless these same lobbyists, chemical company CEOs, and employees all buy farmworker-produced fruits and vegetables grown with toxic pesticides. Or, more probably, they pay a premium for organic produce instead of addressing the problem at the source. As critic

Arianne Burford puts it, "The failure to establish safe guidelines for workers is largely due to a political economy within which the profits of agribusiness outweigh the health and lives of workers" (3). A 2014 article in *The Nation* by Susan Freinkel looks at the challenges faced by those who advocate on behalf of farmworkers, especially in regard to the current regulations of the Environmental Protection Agency: "Our current rules clearly aren't doing the job—they serve the interests of the pesticide industry much better than the interests of children, rural families or public health," says Kristen Shafer, policy director of the Pesticide Action Network. "The system is broken" (Freinkel 18). Despite the many years that have passed since the play's setting in 1988, and the play's world premiere in 1992, both the difficulties endured and the strength in the face of those difficulties displayed by farmworkers continue today.

Heroes and Saints depicts the farmworker community's response to their children's illnesses, deformities, and deaths from cancer at unprecedented rates due to pesticides used by growers in the fields that have also contaminated the community's water supply. While Moraga's *The Last Generation* is about the Chicana/o goodlife writing's conceptualization of land as a member of the community, *Heroes and Saints* shows what happens when twentieth-century U.S. food production ignores this value in pursuit of capital accumulation rather than community sustenance. This was also evident in chapter 4's discussion of migrant farmworker fiction. Moraga especially focuses on the ways that the Chicana/o body experiences a wide range of violence from this capitalist exploitation, which is also a legacy of colonization and modernity. Cerezita is the prime example of this violence, but her sister's dead child, her brother's suppressed homosexuality, and the many other children's deaths in the community are also illustrative. They attest to the way the Chicana/o body time and again lies vulnerable on the frontlines of this struggle to survive and to assert goodlife values in the face of the market-driven abstraction of human labor and the reduction of suffering to statistics. With *Heroes and Saints*, Moraga gives readers a case study of a community that gains strength from protesting its conditions. At the same time, she offers a commentary on the ways that departure from goodlife values—due to a legacy of colonization—creates destruction and suffering for all involved, even to the point of destroying the system that attempts to outgrow its own sustenance.

As an articulate, well-read, and thoughtful survivor of toxic exposure, it is clear why Cerezita constitutes the center of Moraga's play. How better to represent the violence of twentieth-century labor exploitation than a head that has no body with which to labor? Cerezita shows that the abstraction of labor eventually leads to the destruction of the very elements of the system that are necessary for it to function. In actuality, of course, a bodiless head could not survive, but the constant presence of Cerezita on the stage compels the audience to repeatedly confront the exploitation and utter perversion that sustains late twentieth- and early twenty-first-century agriculture. In short, the injustice in the fields is not a sustainable way to live and work. Cerezita is a commanding presence on the stage despite—indeed because of—her condition. She communicates dignity, intelligence, and sensitivity in defiance of her severe physical limitations. She even lives with a small degree of independence, granted her by the "raite" (ride) that her family devises for her to use. The "raite" is a "rolling, tablelike platform . . . automated by a button she operates with her chin" (Moraga, *Heroes* 90). When the audience first meets her, she is a young woman who actively contributes to her family's life and behaves like a typical teenager who dreams of a life beyond her hometown and enacts small rebellions against her mother's strict rules.

In Cerezita, Moraga offers a chilling yet sympathetic character. Still, a disembodied head seems an exaggeration, an extreme. Yet if considered in the context of Chicana/o literary history and Mesoamerican cultures, the disembodied head has antecedents. Most recently, Luis Valdez immortalized a disembodied head when he heard the report that Pancho Villa's body had been disinterred and found to be decapitated. With his play *The Shrunken Head of Pancho Villa* (1964), which includes a character who survives as a bodiless head, Valdez comments on the tensions within Chicana/o identity, especially regarding the connection to Mexico and the temptations to assimilate in the United States. Moraga writes that Valdez's "head character became, for me, a point of departure" (*Heroes* 89). Moraga also cites Olmec head sculptures created in pre-Columbian Mexico as inspirational for her conceptualization of Cerezita's dignified presence. The Olmec head sculptures command respect, with some reaching up to ten feet tall and weighing up to forty tons. Moraga says that "the huge head figures of the pre-Columbian Olmecas are an apt comparison" to Cerezita's solemnity (90). Cerezita's condition references both Chicana/o literary history and pre-Columbian cultures. In this way, we can read Cerezita in continuity with these elements of Chicana/o culture, rather than as an anomaly or a monstrosity. Already at first glance, Moraga challenges readers and audience members to consider the legacy of coloniality in the play's late twentieth-century setting. Cerezita represents not only the injustice of commercial agricultural practices, but also a people's survival and dignity while enduring centuries of colonization.

Since the play's publication in 1992, scholarly readings of *Heroes and Saints* have focused on a range of concerns: disability, the health-care crisis, violence, and death, as well as the expected topics of pesticide use and farmworker justice. Combined, the emphasis of articles on specific aspects of the play and the social issues it tackles provide an overview of the multiple challenges faced in the ongoing decolonization process. In her comprehensive and nuanced study of Moraga's oeuvre, *The Wounded Heart: Writings on Cherríe Moraga*, Yvonne Yarbro-Bejarano underscores the yearning in this play for community by interrogating Chicana identity: "*Heroes* . . . imagine[s] the struggles of the Chicana subject to reconcile her desire for autonomy as well as community, not by passively accepting cultural consensus but by actively negotiating the meanings of cultural values and behaviors" (65).

Yarbro-Bejarano takes particular note of the process of negotiation involved in Chicana identity and the quest for community, observing that this is an ongoing project for Moraga. Yarbro-Bejarano describes how this quest for community involves a differentiation between the experiences of pleasure and joy, which she derives from Cornel West. West describes pleasure as an individual experience, while joy "has to do with love, solidarity, and the struggle for justice" (quoted in Yarbro-Bejarano 65). In her reading of *Heroes*, Yarbro-Bejarano shows how the play depicts the "ethical and epistemological concerns" in the shift from individual to collective and from pleasure to joy. In particular, this play specifies how "the conflicts preventing access to joy are particularly dense in relation to gender roles and sexuality within the family" (66).

This interpretation resonates with the way I see this play lending to an ongoing refinement of Chicana/o goodlife writing. Moraga does more than merely observe the ways that Chicanas/os participate in the values and practices of the goodlife: she also anticipates some of the challenges from within the culture that can stand in the way of truly accomplishing the goodlife, for what is the goodlife without joy? Living ethically with one another and with the planet can certainly ensure our survival, but Moraga asks us to consider what lies beyond mere survival and to strive for joyful communion. Ironically, as she interrogates these possibilities, she draws attention to one of the unique and, to some, disturbing aspects of Mexican and Chicana/o culture: the celebration of death.

Death's significance becomes apparent in the first scene of the play, which consists of actors moving onto the stage with no dialogue at all. "Children wearing calavera masks" enter and hang a child's corpse from a small cross, while Cerezita wheels onto the stage and a helicopter buzzes the field where the cross sits (*Heroes* 92). This opening scene introduces one of the ways in which the community responds to the illnesses and deaths of their children: instead of burying the dead bodies, they hang them on crosses in the pesticide-infested fields. Considered grotesque by some, the town of McLaughlin captures the attention of the media, who send enterprising television reporter Ana Perez, described as "expertly made up" and who betrays a lack of empathy and understanding despite her surname (92). When she interviews the locals on camera, she coarsely remarks that she will "edit . . . out later" a woman's frank description of the town's power dynamics (94). At the end of her segment, Perez enthusiastically remarks that next week's show will take place in San Francisco's Mission District, where they will be celebrating "the Day of the Dead, the Mexican Halloween" (94). Such a facile equivalency between the Day of the Dead and Halloween can only be expressed by one who does not understand the Days of the Dead in anything but a superficial way. Clearly, the children's *calavera* (skull) masks in the first scene evoke the Days of the Dead. Perez's mention of it makes the point more emphatic.

On the Days of the Dead—November 1 and 2—family members create altars for and visit the graves of their loved ones who have passed away. The holiday coincides with All Saints' and All Souls' Days in the Catholic tradition, and families often attend special masses on these dates. But the holiday is recognized as a blend between preconquest practices and beliefs with the postconquest traditions of the Roman Catholic Church. Scholars agree that the holiday shows the Mexican culture's unique relationship with death, although debates continue about the extent to which Mexicans scoff at death entirely.[2] In more recent times, the holiday has become better known and popular outside of Mexico, and its commercial potential is also becoming exploited within Mexico. *Heroes*'s introduction of this holiday sets the tone for an ongoing consideration of death and how the community puts it to use in its protest of their living and working conditions.

The local authorities call the child corpse "crucifixions" a "crime" and launch an investigation to find the guilty parties (Moraga, *Heroes* 93). Given the common practices in late twentieth-century U.S. culture of burying or cremating the dead, this open display of children's corpses can appear a grotesque and extreme measure. Still, the community continues the practice and even Cerezita's baby niece eventually ends up on display. Linda Margarita Greenburg calls the practice the "pedagogy of crucifixion" and points out that it "teaches the audience to re-read the dead children as corporate murder rather than private loss, as active sacrifice rather than victimhood, and finally as social transformation rather than passive necrocitizenship" (165). In the final scene of the play, Cerezita's dead baby niece is hung on a cross while her brother leads the charge to set the fields aflame, and Cerezita sacrifices herself to the cause by wheeling her "raite" into the middle of the fray, presumably dying in the end. This scene shows the community finally coming together as one in the protest, transcending the divisions of gender, sexuality, class, and religion that had previously riven them apart. This final scene also emphasizes, beyond any doubt, the symbolic and literal use of death in this protest. Why must this community resort to the highest sacrifice in order to make its voice heard?

This community, in its display of death, brings attention to the very thing that ultimately drives those who exploit the farmworkers: the fear of death. One of the hallmarks of modernity's progress has been its attempt to control nature, including prolonging life in defiance of death. This has led to such a deeply ingrained fear of death that for one to try to avoid it at all costs remains largely unquestioned. Death, which makes such a vital contribution to the life cycle, finds itself in a decidedly unpopular position in modern and even postmodern culture today. Recognition of and humility before the importance of death would challenge a fundamental value of modernity. The result has been an insidious spread of this ultimately futile defiance of death to all aspects of life today. Some advocate for the possibility of "death with dignity," given our current medical practices of keeping people alive artificially. But this fear touches other aspects of everyday life as well. What is the accumulation of profit or material wealth except an attempt to defy death? If one remained humble in the face of one's mortality, the gathering of outsized material goods would not make sense, in terms of sustainable practices as well as in terms of life span. Why gather what you won't be able to enjoy? Why not share what you have while you are alive instead of keeping it all to yourself in anticipation of living forever? Such accumulation of wealth is often justified in terms of passing it on to the next generations. But *Heroes* asks, what is more important: to pass along material wealth or to pass along a livable planet?

At a rally about access to clean water, the protest leader, Amparo, gives a moving speech to her fellow activists and for Ana Perez's television crew. Amparo argues that the community has been duped by the lure of material comforts—"homes . . . [that] have become prisons" and "water that pours from the sink gots [*sic*] to be boiled three times before it passes your children's lips" (Moraga, *Heroes* 111). She makes the case that they had it better before the signing of contracts and the accumulation of goods, when they were free to live sleeping under the stars and breathing clean air. Reaching the pinnacle of her speech, she asks: "¿Qué significa que the three things in life—el aire, el agua, y la tierra—que we always had enough of, even in our pueblitos en México, ya no tenemos? Sí, parece que tenemos all that we need . . . Pero es mentira" (111).

She asks: what does it mean when the three elements of life—air, water, and earth—which they always enjoyed, even in Mexico, are no longer available to them? She states that it appears that they have all that they need, but then she emphatically tells them that it is a lie. The lure of material conveniences has left them vulnerable to the values and practices that attempt to defy death, and in the meantime the farmworkers themselves have become the canon fodder in this battle. They fall on the frontlines of this attempted defeat of death, with all others benefitting from their sacrifice.

Given these circumstances, it seems only appropriate that this community display its children's mutilated dead bodies to draw attention to its crisis. These "crucifixions" make visible the actual cost of overconsuming middle- and upper-class lifestyles and the pursuit by corporations of profit by any means. The accumulation of wealth remains a respectable goal in the United States. The U.S. criminal justice system never prosecutes exploitations inherent in consumerism and rarely in the case of profiteering. All the while, farmworker children pay the ultimate price for consumers' access to bounteous and beautiful produce. The "crime" of hanging a dead child on a cross pales in comparison. Marilyn Chandler McEntyre argues that this play demands that its audience "take the measure of your privilege. Reckon with your participation in unjust systems of food production, compensation, health care" (98). The community of McLaughlin sees the "crucifixion" of their dead children as a way to honor their lives by giving them a voice in the afterlife protesting the conditions that must change to prevent yet more premature deaths. They make visible the murder committed by the abstraction of labor. For while Moraga shows the Chicana/o understanding of death as part of the overall life cycle, the foreshortening of a young, innocent life violates the integrity of the cycle no less than the artificial extension of life and, all too soon, robs the community of its loved ones and its future. The community protests the unjust deaths that occur in the name of superficial and excessive comforts for others.

THE HUNGRY WOMAN: FINAL LESSONS IN TRANSCENDING POSSESSION

The Hungry Woman: A Mexican Medea, first published in 2000, completes the ecological vision trilogy that *The Last Generation* initiates and *Heroes and Saints* continues. *The Hungry Woman* offers the most difficult aspect of the ecological vision by showing what went wrong with pre-Columbian Aztecs, with ancient Greeks, and with colonial Mexicans and what will go wrong in the future of an imagined ethnic balkanization of the United States. The play invokes legends from all of these cultures and shows how the same thing wrecks them all: the dominance of patriarchal control, hierarchy, and possessiveness. Moraga dwells on what goes wrong in order to keep tragedy from reoccurring, but to do this, she focuses on a disaster that occurs within the community itself. Even so, this play offers a way out. When the land becomes integral to the community, it can overcome these destructive values.

In *The Hungry Woman*, the protagonist Medea winds up poisoning her son Chac Mool. This play's heartbreaking ending brings more attention to the argument Moraga makes in all three ecological vision works: the land is an undeniable member of our community and failing to recognize that puts the entire community at risk. In *The Hungry Woman*, she compares the land to the community's children, and specifically to Medea's son Chac Mool. The murder of Chac Mool shows that when we exploit our land, we murder our own generative forces, just as we would if we had killed our own children. This play does not dwell on how patriarchy and capital that is external to the community exploit the land and the people, such as in *Heroes and Saints*. Instead, Moraga warns about the way that the Chicana/o culture itself assimilates exploitative values.

At the start of the play, Medea finds herself living in a mental institution just outside the border of Phoenix, Arizona, an environmental "wasteland where yerbas grow bitter for lack of water" (Moraga, *Hungry Woman* 15). We later discover that she is in the mental institution because she poisoned her teenage son Chac Mool. But before this final tragedy occurs, we learn that Aztlán's leaders exiled Medea, who was a prominent warrior in the rebellion, after she falls in love with a woman. They make no space for queers in the future Chicano homeland. Medea, her lover Luna, her mother Mama Sal, and five-year-old Chac Mool leave Aztlán for the "no man's land" of Phoenix to live with the others who are unwanted by the new regimes. By the beginning of the play, Medea and Luna have a troubled relationship, while at the same time Chac Mool's father Jasón demands the return of Chac Mool to Aztlán. After Chac Mool agrees to abandon her and return to Aztlán, Medea poisons him. She gives up on the idea that he can become a man who respects women. If he were to live under the regime of Aztlán—a regime that has seen fit to exile her and Luna—that would be impossible. Her desperate deed is why she ends up in the mental ward. In a style that challenges the audience to master anti-chronology, the play gives readers a series of flashbacks to attempt to explain Medea's madness.

Moraga sheds light on the reason why the Mexican Medea and other supposedly "badly behaved" women from mythology and folklore sacrifice their children. They have been attempting to draw attention to the destructiveness of exploitative values and behaviors. When Medea sees that her son is on the brink of becoming a man, she ends his life because she understands that his maturity means betrayal. No role models exist that teach him to value and respect women. Medea tells Chac Mool's father, Jasón: "The man I wish my son to be does not exist, must be invented. He will invent himself if he must, but he will not grow up to learn betrayal from your example" (69). At this point in the play Medea hopes she can offer Chac Mool the space to invent a self that does not betray the women in his life. She also adds, "Betrayal occurs when a boy grows into a man and sees his mother as a woman for the first time. A woman. A thing. A creature to be controlled" (70). Like the Greek Medea, who sacrifices her children, like Coyolxauhqui, who tries to kill her own mother, like La Llorona, who drowns her children, the Mexican Medea prematurely ends her son's life so he does not live to reproduce the same exploitative system of control and possession.

Although the play brings together into a single narrative a variety of myths and legends from different time periods and traditions, they all hold the killing of closely related family members in common. In these legends the play reveals its hope for the community's redemption. Readers familiar with Greek mythology will immediately recognize the antagonism between Medea and

Jason. Euripedes's Medea and Jason battle over love, land, and children in ways similar to Moraga's Medea and Jasón. Both Medeas also kill their children. The parallel ends with Medea's relationship with Luna, a young sculptor she falls in love with in Aztlán, and who recalls a different—but no less significant—mythology for the readers. In the prelude of act 2, the stage directions state that "LUNA *appears as* COYOLXAUHQUI" (55) at the same time that the legend of Coyolxauhqui is being read to the audience. Those familiar with this Aztec legend relating to the origin of the sun, moon, and stars know that Coyolxauhqui and her many siblings attempt to kill their mother, Coatlicue, when Coyolxauhqui discovers that their mother will give birth to Huitzilopochtli, the god of war. But the god of war emerges from the womb brandishing weapons. He protects his mother and saves himself by dismembering his sister Coyolxauhqui. As the legend is read, it is also being dramatized on stage. After Huitzilopochtli triumphs, he asks, "Is this my sister's moonface I hold bleeding between my hands? I exile you foreign and female into that vast hole of darkness that is your home" and then the stage directions state "*He tosses 'the head' into the heavens. They all watch the moon rise into the night sky*" (56). Thus goes the Aztec origin story for the moon (Coyolxauhqui), the stars (Coyolxauhqui's many defeated siblings), and the sun (Huitzilopochtli), who refuses to share the sky with his siblings.

In this play, the Coyolxauhqui legend makes explicit the regard in which patriarchy holds women: as "foreign" and at home in "darkness" (56). This, in essence, expresses the views that Medea fears Chac Mool will acquire if he moves back to Aztlán. Of course, Coyolxauhqui/Luna embraces the power of darkness and becomes the light in the night sky, upsetting patriarchy's attempt to relegate the female to nothingness. Critic Alicia Arrizon states, "According to Moraga . . . Coyolxauhqui's disobedience functions in the play as an attack on the larger frame of patriarchy" (48). Given that the play offers so strong a condemnation of patriarchy that the sacrifice of Chac Mool is warranted, does it offer an alternative?

Three more legends figure into this play's action, and together they indicate a possible alternative community structure, one that both resists hierarchy and follows ecological cycles. The three legends that gesture toward an alternative center on these figures: Chac Mool, an Aztec rain god; the Hungry Woman, an Aztec creation myth; and La Llorona, a colonial-era tale that Mexican Americans still tell today. The role of Chac Mool is the only one played by a male in the entire production (Jasón is played by a female actor). Medea gives her son the name Chac Mool only when they leave Aztlán. His father had originally named him Adolfo, but Medea reveals that she never wanted to call him that and instead gave him an indigenous name that could make him proud, and he identifies with it so much that he gets it tattooed on his arm "so [he] won't forget" (Moraga, *Hungry Woman* 71). After Chac Mool gets the tattoo, he asks his mother "he's a warrior, right?" (29). When Medea reveals that she has named him after "a fallen warrior," Chac Mool is disappointed—because he represents "someone who didn't win" (29). Medea tells him that "winning's not the point" (29). The evocation of the mythic Chac Mool and therefore the significance of the name to the meaning of the play becomes difficult to parse when one looks into this name's background.

Chac Mool exists as a rain god in Mayan mythology and under a different name, Tlaloc, he is also a rain god in Aztec mythology. Given the influence of rain on agriculture and sustenance for a people's survival, a rain god holds considerable sway. However, Medea does not emphasize this particular background for her son's name. Experts on the statuary of the Mayan city Chichen Itza explain that a *chac mool* is a three-dimensional sculpture depicting a reclining figure found in pre-Columbian Mesoamerican art. The reclining position and the frontal face details indicate a captive or submissive figure. Medea describes the chac mool after whom her son is named to be more like a fallen warrior, which indicates that she references this submissive sculpture more than she invokes the rain god association of Chac Mool. Perhaps she expected to communicate something specific to her son with this name: that "winning's not the point" (29). In the past, she told him that a chac mool was a warrior. In truth, a chac mool was a warrior that had been captured, and therefore fallen.

What point does Medea want to make clear to Chac Mool by finally telling him this about his name? Medea's other name, the Hungry Woman, can help answer this question. This Aztec creation myth shows how fighting for survival does not always occur on the battlefield and that life cannot exist without relinquishing control and allowing for death. The Hungry Woman legend makes clear that including land as part of the community keeps this knowledge front and center.

The creation story "The Hungry Woman" provides part of the title for the play and is told in the course of the play, as well. Luna recounts the story for the audience after she and Medea have argued and made up. They begin to make love, and Luna tells the story at the same time that the stage directions indicate, "LUNA *makes love to* MEDEA *with her mouth*" (44). The hunger that Luna satisfies as she tells the story further illustrates the generative power of the female that "The Hungry Woman" illustrates: this goddess embodies a passionate life force. "The Hungry Woman," in brief, is the story of a woman who, at a time before the Earth was created, disturbed the gods with her constant crying. She was wailing because the mouths at each of her major joints (wrists, elbows, knees, ankles) were not being fed. The male gods Tezcatlipoca and Huitzilopochtli refuse to feed her. To quiet her, they attempt to drown her, but in the process they accidently split her in two. She cries even more. Feeling remorseful for their violence, the gods decide to comfort her by putting her body to the redemptive use of creating the Earth. They made "grass and flowers out of her skin" and from "her hair they made forests, from her eyes, pools and springs, from her shoulders, mountains, and from her nose, valleys" (Bierhorst 25). But her hunger and her moaning continue. Now, when anything dies, she consumes it, and still she cries. "Sometimes at night, when the wind blows, you can hear her crying for food" (25).

More than simply a Mother Earth figure, the Hungry Woman ("Cipactli" in Nahuatl) embodies the ecological cycle of life, death, and regeneration. She not only consumes life (demands feeding, demands death for her survival), but her body also provides the basis for all life (as the flora, waters, and land formations of Earth). The gods, in their attempt to control the feminine, instead make this goddess the font of life on Earth. The centrality of this tale to the events of this play shows Moraga's emphasis on the power of the feminine despite patriarchal attempts to overpower it, as well as demonstrates the importance of destruction to the generation of life. "The Hungry Woman" gives readers the opportunity to feel empathy for this creature, which is the land itself, and the play reveals her pain as well as her power. Readers can identify with the land and suffer alongside her. In short, Moraga challenges readers to consider the land to be a family member as much as one considers a child or a parent to be.

When Medea kills Chac Mool, readers are meant to understand that she simply reifies the spiritual poisoning that she anticipates he will ingest when he moves to Aztlán. The land of Aztlán itself suffers from the same poisonous exploitation under the new regime as it did under the old. It is a harsh sentence that she imposes on her own son, but in two different instances the play signals that he ultimately understands. In the first instance, Chac Mool invokes another tale that provides context for this play: La Llorona. Akin to the Hungry Woman, La Llorona is also known for her *llanto* (weeping). The last line of the Hungry Woman legend brings La Llorona to mind: "Sometimes por la noche, when the wind blows, you can hear her crying for food" (Moraga, *Hungry Woman* 45). The legend of La Llorona can generally be summarized as "the bogeywoman of Mexican and Chicano folklore said to have murdered her children and who is doomed to wander for all of eternity as punishment for her crime. She wails at night as she wanders dark roads searching for her children or any other potential victims" (Perez 154).

The reason given for her murders always has to do with the betrayal she suffers at the hands of a man. Some scholars trace this story back to the colonial-era relationships between Spanish soldiers and indigenous women who were abandoned by their lovers when their Spanish wives arrived in the New World (Perez 156). In this way, the tale of La Llorona expresses centuries of pain and heartache. Every generation of Mexicans and Mexican Americans learns of this love gone terribly wrong, and sympathy rarely is given to the woman, who is considered contemptible for her actions.

However, Chicana feminist writers and scholars have recovered the indigenous woman's side of the story and argue that "the woman's actions are a direct response to Spanish cultural, economic, or social oppression. As such, [La Llorona] emerges as a figure of resistance rather than a complete cultural villain" (Perez 156). In this play, Chac Mool reveals his own openness to understanding La Llorona's experience. He recalls hearing her wailing near a canyon when he still lived in Aztlán as a child. His grandmother asks him if he was ever afraid of La Llorona, and Chac Mool replies that he felt empathy, not fear: "Her voice was inside the wind . . . I felt like she was telling me her side of the story, like I was the only one that heard it like that" (Moraga, *Hungry Woman* 38). Does his understanding of La Llorona mean that he later understands his own mother's sacrifice of him? This play indicates that he does.

At the end of the play, when Medea is living a tortured existence in the mental hospital, Chac Mool's spirit visits her and helps her transition to the next world. If he did not understand or forgive his mother's actions, he would not help relieve her pain and ask her to join him. When she sees Chac Mool again, she knows that he must be an apparition and she tells him that she is insane. He replies, "I don't know" (98), revealing his lack of confidence in scientific diagnoses. Instead, he prefers to care for her himself. He tells her, "You're going home" and "Here, drink this. It'll help you sleep" (99). With this, the play ends, and Chac Mool holds Medea in a pietà image, but of course the gender and parental roles are reversed. Given Chac Mool's empathy with La Llorona and his rescue of Medea at the end of the play, readers can see that he also understands why Medea poisoned him.

One last connection between the Hungry Woman and La Llorona helps clarify Moraga's correlation between Chac Mool and the land. In La Llorona's legend, she is relegated to wandering along waterways and weeping as her punishment for drowning her children, with witnesses often hearing her wailing at dusk and nighttime at these locales. The Hungry Woman is also associated with waterways, given that the "pools and springs" of the Earth are made "from her eyes" (Bierhorst 25). Chicana feminist scholar Domino Renee Perez observes "the symbolic significance of the water, as represented through La Llorona's tears or the rivers, lakes, and streams where she is said to wander, allows for the possibility of rebirth and/or baptism" (156). Moraga invokes these two legendary figures that strongly associate with water and thus allow for the possibility of a new life or a rebirth for the community. Perhaps the rebirth will not occur at the present time, but at least the possibility remains open. Rebirth and new life can occur at a time when the community can recognize all its members as fully deserving dignity and respect, including the land, or the Hungry Woman.

Critic Patricia Ybarra calls the events that follow the revolution as depicted in *The Hungry Woman* "a 'productive' type of failure for rethinking revolution in the Americas" (Ybarra 64). She adds that "*The Hungry Woman* depends on a conception of futurity shared by many Mexican indigenous tragedies, one that places the fruition of present bodily sacrifice in a promised but not yet realized solidarity" (67). My reading of this play resonates with Patricia Ybarra's perceptive reading of the futurity evident in Moraga's ending. I, too, see how Moraga shows that once the revolution's goals are achieved, exploitative and hierarchical values still survive within the community. These persistent values infiltrate the revolution and destroy its all-too-brief unity. Moraga identifies specific intracultural tensions that she notes in other works: Chicana/o homophobia and patriarchy. But even as *The Hungry Woman* depicts tragic events, a closer look at the five legends that impact the telling of this play's events betray utopian longings for the community. *The Hungry Woman* interrogates the possibility of joy within the community by using deep history to show alternative genealogies that create possibilities for the present and future. Death is once again just a part of life rather than something to paralyze us with fear.

CONCLUSION

Moraga wrote other works between *The Last Generation*, *Heroes and Saints*, and *The Hungry Woman*. However, I bring these three works into dialogue across the decade that spans their publication dates because of their common exploration of the idea of transcending possession in the community's search for joy. If *The Last Generation* challenges the community to work together to avoid its own extinction, then *Heroes and Saints* depicts the difficulties the community faces as its members band together against oppression, and *The Hungry Woman* shows the disappointing outcomes after the longed-for revolution if the poisons of patriarchy, homophobia, and possessiveness are not purged. In all three works, Moraga consistently offers a view of community coalescence through the release of control over each other, the natural environment, and life itself.

This understanding of these works comes at a much-needed time when we do not know how to relate to an environment in ever-deepening crisis. Environmental philosopher Catriona Mortimer-Sandilands comments that there is "certainly no systemic recognition that [environmental] loss might be (literally) earthshattering for many people, akin to the death of a lover, a parent, a child" (339). She asks the intriguing question, "How does one grieve in a context in which the significance, the density, and even the existence of loss is unrecognized?" (339). For Mortimer-Sandilands, "queer melancholia," which theorizes how to mourn a connection that goes largely unrecognized and unrespected, helps us understand how to mourn the gradual and ongoing loss of a human-sustaining environment. I would add that Moraga's focus on death and the embodiment of land in the form of the Hungry Woman in these three works helps address the lack of mourning ritual or even understanding that exists today for nature. Modernity/coloniality has not developed a capacity to mourn the loss of a human-sustaining environment, because it has not recognized nature's agency and subjectivity. Moraga's and others' goodlife writing documented in this book do allow for nature's agency and subjectivity. Moraga's use of death in particular as a fulcrum for critiquing modernity/coloniality brings attention not only to modernity/coloniality's exploitations, but also to its weaknesses in the face of natural processes that sustain human life. Decolonial environmentalisms, like those found in these three works, honor the place of death in the process of life, and they undo the power that modernity/coloniality wields through false promises to shield us from death. Ironically, this attention to death helps us mourn the disappearance of nature and helps us to better see what we need to do—honor human-sustaining, ecologically sensitive good-life—in order to survive and flourish.

CONCLUSION

Decolonized Environmentalisms for the Twenty-First Century

IN THE TIME it has taken me to research and write this book, I have seen the interest in and documentation of Latina/o and Chicana/o environmentalisms grow. This field of study has gone from being a marginalized intellectual inquiry to building momentum toward a fully fledged scholarly field and activist arena. Latina/o and Chicana/o environmental activists enter the debates on climate change and eco-justice with power and commitment. The genealogy I have traced in this book can offer at least partial insight into how Latina/o and Chicana/o scholars, writers, and activists speak to environmental issues from a strong position, with confidence and authoritative knowledge. Even though we may not have been recognized as environmentalists, and much less as members of the green movement in the United States, we still lay claim to at least a century and a half of contribution to the human relationship with the natural environment. In this conclusion, I document just a few of the most recent cultural productions that emerge when Chicana/o writers, scholars, and activists make known their commitments that can almost invariably be traced back to an environmental issue. In particular, I note that each instance of environmental awareness links up with social justice endeavors, making clear how the Chicana/o vision connects with our ongoing struggles to create space for our ways of being and knowing. As Ontiveros states, "Unlike mainstream environmentalism, which tends to draw a bright line between society and nature, Chicano/a environmentalism sees social injustice and environmental damage as part of the same structures of inequality and indifference" (130). Decolonized environmentalisms for the twenty-first century interconnect a network of critique and activisms that overall call for the advancement of the goodlife.

Specifically, this conclusion will discuss two political phenomena in the United States today—the attack on immigrants and the dismantling of ethnic studies—and show how Chicana/o writers' and activists' responses are informed not only by social justice but also by environmental logics. In 2010 the Arizona legislature passed two controversial laws that set a new national standard for anti-immigrant nativism, which also included a linked strategy to discontinue ethnic studies. The first law, Senate Bill 1070, drew immediate attention to Arizona for its criminalization of not only undocumented workers, but also anyone who hired or assisted them. Moreover, it gave local authorities the ability to demand proof of legal status from anyone at any time, as determined by law enforcement officials. Critics immediately interpreted this as a legalization of racial profiling in Arizona, especially of the Latina/o community. S.B. 1070 quickly became known as the "Show Me Your Papers Law." In 2012 the U.S. Supreme Court upheld this law's requirement of proof of citizenship, but struck down aspects of the law that allowed the police to incarcerate suspects for long periods of time. House Bill 2281, passed in the same legislative session as S.B. 1070, created less immediate response, but it eventually became known as the "Anti-Ethnic Studies Law." It decreed the withdrawal of state funding from public schools where courses "promote resentment toward a race or class of people," that "advocate ethnic solidarity instead of the treatment of pupils as individuals," or that are "designed primarily for pupils of a particular ethnic group" (Arizona H.B. 2281). This law is still in the appeals process five years later.

Together these bills represent a two-pronged assault on Latinidad—Latina/o being and knowing—in Arizona, which some have called an attempted ethnocide. By casting such a wide net with the anti-immigrant legislation, S.B. 1070 criminalized the Latina/o-appearing body and anyone allied with a Latina/o or other immigrant-appearing body. By outlawing ethnic solidarity and the teaching of historical events that could generate resentment, such as Jim Crow segregation or the American imperial takeover of half of Mexico's territory, H.B. 2281 censored the teaching of Latina/o and other ethnic minorities' experiences and ways of being and knowing. At the same time, it keeps the dominant population from learning about the hypocrisy of its instituted values and power structure. Of course, an irony of H.B. 2281 is that it outlaws the design of courses with a particular ethnic group in mind without recognizing that the longstanding and mainstream curricula continue to overwhelmingly privilege the Anglo-American population. But what does any of this have to do with environmental issues?

In the spring 2009 issue of the journal *Environmental Ethics*, philosopher Phillip Cafaro and wildlife biologist Winthrop Staples III made the "environmental argument" against immigration. In 2015 Cafaro published his book *How Many Is Too Many?: The Progressive Argument Against Immigration into the United States*. The publication of this scholarship in an environmental context and the passing of S.B. 1070 in Arizona just a year later is no coincidence. Nativist anti-immigrant sentiment has been building in the United States, drawing upon any resource to bolster their case. On the June 19, 2015, episode of *Real Time with Bill Maher*, Ann Coulter, author of *¡Adios, America!: The Left's Plan to Turn Our Country into a Third World Hellhole*, commented that immigrants litter "like mad" and that they "are dumping pesticides in the Redwood Forest" (Maher 2015). She added that "they're at the slash and burn stage of capitalism" and that "they have no concept of litter." When, on one hand, liberals utter surprise that Latinas/os care about climate change (see introduction), and, on the other hand, reactionary nativists paint immigrants as lacking a capacity to understand the fairly simple concept of litter, one can see how the mainstream might expect little environmental knowledge to emerge from this population. However, as this book shows, Latinas/os (including immigrants) offer a depth of knowledge and experience about the human-nature relationship. This book has only been about one group of Latinas/os—Mexican Americans and Chicanas/os—but Latinidad encompasses a vast and complex population with many yet unappreciated values, experiences, and epistemologies.

Cafaro and Staples's journal article represents the most seemingly objective attempt to link an anti-immigration stance with upholding what they consider a morally correct environmental practice. Even so, the argument breaks down rapidly. They reason that the United States makes the biggest ecological footprint on the planet. Given this fact, more growth in the U.S. population translates to an even bigger U.S. ecological footprint. Cafaro and Staples present many graphs and statistics to demonstrate their point. This makes sense, and is indeed the case, but this narrow linking of the U.S. ecological footprint to an immigrant population alone leaves a great deal out of the discussion. Why not target the U.S. hyperconsumerist lifestyle? Immigrants and minority U.S.

ethnic communities can help turn the consumerist tide, despite Coulter's cynicism. Those living on the edge socially and economically often have thrifty habits out of necessity, but sometimes also out of cultural legacy. Add to that the link to oppressed cultures that have resisted and adapted modernity for generations, and there is a lot from which to learn. But Cafaro and Staples, not to mention Coulter, never consider this possibility. They simply mention that lowering consumption rates can be an option, but it is too big a trend to reverse at this point: "re-engineering the world's largest economy and changing the consumption patterns of hundreds of millions of people are immense undertakings that will be difficult, expensive and (we may assume) only partly successful" (Cafaro and Staples 12). They also only briefly consider the idea that U.S. policies' impacts on other nations' environments and economies might be motivating immigration to the United States. In their concluding policy proposal they submit a three-pronged approach: cut legal immigration, reduce illegal immigration by creating laws like S.B. 1070 and enforcing them, and rework trade agreements while increasing foreign aid.

The last approach, which involves changing U.S. foreign policy, treats other nations as trading partners and natural resources rather than as autonomous regions. Curiously, Cafaro and Staples treat the United States as somehow environmentally autonomous when they propose that limiting its rates of immigration will help the global environment. But we live in a deeply linked planet—ecologically, politically, socially, and economically. Reducing immigration would be a mere Band-Aid fix. So I agree with the strategy to rework trade deals and increase foreign aid, for starters, but I wonder why Cafaro and Staples do not shy away from these challenges the same way they do from the prospect of reducing consumption within U.S. boundaries. Foreign policy changes can be just as tall an order as changing U.S. rates of consumption. One is more inward looking than outward bound, and I would think that the domestic approach might better appeal to nativists. But of course the desire to keep out "the foreign" overrides reason at times. It has become too easy to join the bandwagon of anti-immigrant sentiment in the United States. What gets left out of this discussion is what nativists perceive as "foreign" can bring to the environmental challenges overall. Environmental nativist rhetoric tragically underestimates whole classes of peoples and is blind to the diversity within our borders. It also insults the very idea of ecological connectivity, as if maintaining a political border will make a difference when the entire planet's climate is in crisis. We can best benefit by examining what drives the destruction—such as exploitative

FIGURE 5. Still shot from *El Librotraficante* video. Tony Diaz, 2012. Shot by Bryan Parras.

trade deals and cheap imports—than trying to bolster false boundaries. As tall an order as it might be, the United States, and the planet, would most benefit from dismantling (not "reworking") trade deals such as NAFTA, which wreak havoc with national economies and run roughshod over environmental protections. This would give potential immigrants to the United States the means to stay home, which I expect they would prefer to living in the United States, where they encounter racism and oppression. Taking a deeper look at the reasons for immigration would better serve the nativist interests. At the same time, the United States would better invest its energies in reducing rates of consumption rather than building taller walls and flying more drones over the U.S.-Mexico borderlands, wasting resources and disrupting natural migrations of all sorts.

A few of the Chicana/o responses to narrow-minded nativism can give us an idea of what decolonized environmentalisms might look like for the twenty-first century. I first give a portrait of the creative and dynamic Librotraficante movement that was born in response to H.B. 2281. Then I briefly discuss two books: *The Guardians*, a 2007 novel by Ana Castillo, and *The Devil's Highway*, a 2004 nonfiction book about the tragic death of fourteen migrants in the harsh Sonoran Desert. I conclude with a quick survey of the mix of stories from the nationally syndicated radio program *LatinoUSA*'s series called "RadioNature," which reports on a broad array of topics related to Latina/o life and the natural environment. The RadioNature reports can offer an impression of how Latina/o popular culture responds to conventional expectations of what environmentalism looks like. On the whole, the works that I discuss in this chapter—grassroots activism, two novels, and radio reports—will not exclusively engage environmental issues. But that fact speaks to my point in this chapter: decolonized environmentalisms for the twenty-first century understand that isolating the human-nature relation from social justice concerns is to our detriment.

LOS LIBROTRAFICANTES: THE BOOK SMUGGLERS AND HOW MUCH "WE NEED MORE WORDS"

In the spring of 2012, Houston-based literary impresario Tony Diaz stepped in front of a video camera and forever changed the war on ethnic studies by launching a creative and engaging offensive. In the video, Diaz wears dark sunglasses, a brown leather jacket, and a tough guy mug, and tells viewers the story of Arizona's attack on ethnic studies. He takes particular exception to the idea that certain books could no longer stay on the shelves in Arizona schools. He volunteers to educate Arizona legislators with three new vocabulary words: "Librotraficante," "wet-book," and "dime-book." At this point, the viewer clearly understands the parodic tone of the video, and can listen good-humoredly as Diaz defines these three terms. A "Librotraficante" is someone who smuggles banned books back into Arizona. Diaz tells viewers to go to the website www.librotraficante.com to get involved with the caravan to Tucson during Spring Break 2012, which turned out to be a real caravan of busses full of books and activists, and *not* part of the joke. A "wet-book" is a book smuggled back into Arizona by the Spring Break 2012 caravan for use in "underground classes to illegally conduct Latino literary studies." A "dime-book" is a paperback once worth ten dollars, but now invaluable "thanks to your fascist laws, Arizona!" Diaz says. In a final commentary, Diaz states, "We're throwing the book at you, Arizona!" The video lasts a mere one minute and forty-one seconds, but it started a movement that is still going strong today, working to protect dignity and respect for Latinas/os across the nation.

The Librotraficantes were part of a movement that resulted in a big impact on high schools today. According to a July 2015 article in *The Atlantic*, "Mexican

American studies has spread to high schools at a rate that no one could have imagined before Arizona banned the class in 2010" (Phippen). Instead of succeeding in outlawing ethnic studies—and Mexican American studies in particular—Arizona's legislators' strategy backfired on them and resulted in many more programs in a wide array of locales. According to Diaz, H.B. 2281 "sped up the evolution [of Mexican American studies] by about 25 years" (Phippen). Any attempt to pass similar laws in other states gets "the book thrown" at them by the Librotraficante network. In 2013, Texas Republican state senator Dan Patrick and Texas Republican state representative Giovanni Caprioglione each introduced legislation that would no longer allow ethnic American history courses or women's history courses to count toward requirements on college degrees. Combined, Patrick's S.B. 1128 and Capriglione's H.B. 1938 would have effectively gutted enrollments and therefore funding for these programs and departments. The Librotraficantes took immediate action against these legislations and the bills did not pass.

Clearly, the Librotraficante Spring Break 2012 caravan did more than dramatically smuggle books into Tucson. They simultaneously inhabited physical and metaphorical space throughout the country to ensure that nativist attacks on ethnic studies looked ridiculous and ineffective. The caravan organized an array of events in Houston, San Antonio, El Paso, Albuquerque, and Tucson. The Librotraficante events included the creation of "underground libraries" for the local communities and boisterous rallies, such as a poetry reading by Lorna Dee Cervantes in front of the Alamo in San Antonio, Texas. The stop in New Mexico included readings by Rudolfo Anaya, Luis Alberto Urrea, and Jimmy Santiago Baca. Every stop drew crowds of Latinas/os, making a clear point that the communities across the Southwest supported this work. Of course, one cannot help but notice that the physical space covered by the caravan spans a great deal of the territory originally seized from Mexico by the United States. The caravan covered the lands from southwest Texas, where Houston sits nearly at the edge of the Gulf of Mexico, west to Tucson, Arizona, nearing the southern U.S. border. They also called the Arizona legislature on its absurd attempt to criminalize an ethnic identity: the Librotraficantes created a parody of this view by depicting literary activists as scofflaw smugglers, playing up a tough attitude and "peddling" restricted knowledge to the young people they encountered. In a more serious moment, Diaz observes, "It's clear to me that our intellectual advancement is a threat to some people, because they tried to make it illegal" (Phippen). Diaz has known the precarious position of Latina/o intellectual advancement for some time, and his literary advocacy for over fifteen years put him in the right position to call upon a network of writers and volunteers to respond immediately to Arizona's attack on Mexican American studies. He and his network offer an exemplary model of grassroots activism for the twenty-first century that offers hope for creating change in social justice arenas, including environmental injustice.

Diaz is a novelist in his own right and a tireless advocate for Latina/o literature. He has led the nonprofit group, Nuestra Palabra: Latino Writers Having Their Say, in Houston, Texas, since 1998. Nuestra Palabra (NP; Our Word) started out organizing literary showcases that gave community members the opportunity to share their work on stage—poetry, fiction, nonfiction, drama, and music. NP also invited renowned writers for community readings, so Houston had the chance to see writers such as Jimmy Santiago Baca, Dagoberto Gilb, and Sandra Cisneros in events organized by up-and-coming writers and activists such as Diaz himself, Alvaro Saar Rios, Liana Lopez, Bryan Parras, Gwendolyn Zepeda, and many others. Rios is now a professional playwright and assistant professor at the University of Wisconsin–Milwaukee, Zepeda was appointed Houston's first poet laureate in 2013, and Lopez and Parras have expanded their organizing efforts on behalf of ethnic studies and environmental justice. NP also started hosting a weekly radio program that still plays on the local Pacifica public radio station, KPFT 90.1 FM. Additionally, NP organized several successful book fairs called the Latino Book and Family Festival, which drew tens of thousands to the weekend of readings, seminars, booths, and events in downtown Houston. NP not only promotes literary advocacy in its public forums, but it also grooms generations of leaders who emerge from volunteering with priceless skills and experience.

As a graduate student in Houston during the early 2000s, I was privileged to volunteer with NP and join the audience for the literary showcases, book fairs, and radio shows. I also learned early lessons in the synergy between literary advocacy, social/environmental justice, and literary studies. I was enrolled in a graduate program in English, where I had a wonderful mentor and dissertation director in José Aranda, but he was the only professor who specialized in Chicana/o literature on the faculty. NP became an integral part of my experience and learning during those years. As a volunteer with NP, I not only read and wrote about my favorite authors, but I went out to dinner with them and drove them to the airport, too, after I heard them read for enthusiastic audiences full of my fellow Latina/o community members. My informal conversations with Chicana/o authors such as Demetria Martínez, Tammy Gómez, Dagoberto Gilb, Sandra Cisneros, Sergio Troncoso, José Angel Gutierrez, Herbert Siguenza, and Richard Montoya taught me so much more about the dimensions of the writing craft, the role of politics in creativity, and the priority of community loyalty than I could have learned from books alone. After I left Houston with my degree to pursue my career, I missed my days with NP, but I knew both that their work would carry on and that NP had made a lasting impact on deepening my commitment to Latina/o literary arts. Although NP's work is not explicitly environmental, the group works tirelessly to promote Latina/o literature and to create spaces where Latinas/os showcase their creativity and sustain goodlife values. I include their activism here to show the innovative and rapid grassroots response to the values and actions of nativists such as Cafaro, Staples, and the Arizona and Texas legislators.

Given my past experiences, I was not surprised when a YouTube video starring NP founder Tony Diaz, shot by Bryan Parras, crossed my screen in the spring of 2012. By using humor, the Librotraficantes changed the conversation in their favor, clearly showing the generosity inherent in Latina/o studies and ethnic studies in general. As Linda Martín-Alcoff observes in a 2012 *New York Times* op-ed, the Arizona legislators' opposition to ethnic studies emerges out of fear: "Their real concern, as stated in the bill, was about 'solidarity' and 'resentment.' They are scared of a curriculum that might foment an anti-white sentiment among impoverished populations of Mexican and Central American kids . . . H.B. 2281 is an attempt to stem the tide of Latino political integration as full participants, a development that may well change the color of 'Americanness.'" The Librotraficante strategy shows the establishment that Latinas/os are not "out to get" anyone, but only interested in sustaining and sharing our knowledge and experience. In fact, ethnic studies programs in schools across the United States are proven to increase graduation rates across the board. According to Phippen's *Atlantic* piece, "the University of Arizona [published] a report that found offering Mexican American studies increased graduation rates, grades, and college enrollment." Latina/o studies programs in particular help educate our future generations—from all ethnic backgrounds—about a population that is well on its way to demographically transforming this nation.

In a 2010 *New York Times* op-ed, Stanley Fish, who must be commended for his quick response to the passage of this legislation, comments that H.B. 2281 reveals a "clash between two bad paradigms." On one hand, Fish characterizes the pro–ethnic studies camp as advocating a learning environment where "students are to be brought to see what the prevailing orthodoxy labors to occlude so that they can join the effort to topple it." On the other hand, Fish evaluates the anti-ethnic studies camp as using strategies where the "remedy is worse than the disease." He adds that "the authors of House Bill 2281 don't want students to learn about the ethic of treating people equally; they want them to believe in it (as you might believe in the resurrection), and therefore to believe, as they do, that those who interrogate it and show how it has sometimes been invoked in the service of nefarious purposes must be banished from public education" (Fish). Although I am sympathetic with how Fish interprets the lack of intellectual rigor that goes into the anti–ethnic studies sentiments, his reading of the pro–ethnic studies camp underestimates and oversimplifies what goes on in our classrooms.

Far from a focus on toppling "the prevailing orthodoxy," ethnic studies courses labor to recover and sustain knowledge apart from the mainstream. Ethnic studies programs and departments are not concerned with "the prevailing orthodoxy" as much as they are centered on recovering and retaining marginalized voices. If "the prevailing orthodoxy" then finds itself in a position where it can no longer sustain itself alongside ethnic studies, then the fault lies more with the lack of depth and consistency of the mainstream than with an assault from ethnic studies. Time and again, the values I find in Chicana/o texts urge us to reject privilege in favor of distributive power. Any strategy that attempts to wrest dominance from the mainstream and reproduce the same hierarchy spells doom in Chicana/o cultural production. These are the lessons we teach in our classes and that resonate with the goodlife values of dignity, respect, simplicity, sustenance, and transcending possession.

Two of the leaders, with Diaz, of Nuestra Palabra and the Librotraficante movement also figure at the head of environmental justice activism in the Houston area, and thus offer an opportunity to see how these social justice arenas interconnect and overlap. Indeed, their involvement in both realms blurs their boundaries in intriguing ways. Bryan Parras and Liana Lopez have played leading roles in actions protesting toxic environments, including the Keystone XL pipeline. The pipeline's proposed terminus is the oil refineries of the East Houston shipping channel area, where Parras lives. Parras is cofounder, with his father Juan Parras, of the Texas Environmental Justice Advocacy Services group (TEJAS). Lopez works with TEJAS, together with Parras, in leading workshops that put the tools to document environmental injustice in the hands of community members, including photography and videography.

In an interview published in *The Nation* in July 2013, Parras speaks to the ambiguity of environmentalism as an identity for Latina/o communities at the same time that he connects environmental advocacy to other social justice issues. First of all, the interviewer, Wen Stephenson, asks Parras if he identifies as an environmentalist. This is an interesting question seldom asked of white activists, as if this identity chafes with people of color. Parras pushes back at the question a bit, stating that he is willing to "let folks call me that . . . if it's the only way of describing myself that makes sense" (Stephenson). Parras explains that "we need more words" for the kind of work that he and TEJAS are doing. His sentiment resonates with me and represents the impetus for this book's attempt to name what Mexican American and Chicana/o creativity brings to the environmental conversation.

Parras's remarkable sensitivity to the cultural dimension of environmental work connects his Latino literary advocacy with his environmental justice organizing. While he has emerged as a leader in both realms, the media coverage of these events never explicitly links these two areas of activism. But for Parras, the connection seems clear. He states that communities who organize against environmental injustice also make connections to other injustices:

> It's important to understand that it's a dynamic situation, for the people there. And a lot of the injustices are not only environmental-related, in the sense that most people think of the environment. So, chemicals, yes. But police brutality? No. Drugs—you know, other kinds of chemicals? No. Lack of education opportunities? No, people don't think of those things. So it's a dynamic situation that allows these communities to exist in the first place, and then on top of that you have these environmental, toxic exposures that make it even more difficult to get out. (Stephenson)

Parras underlines that for low-income communities and communities of color, "environment" is much more broadly defined, including educational opportunities. His statement weaves together the NP/Librotraficantes with environmental justice advocacies. Access to educational opportunities definitely falls within the realm of environmental justice for Latina/o communities. In April 2014, the Librotraficantes, along with other activists in Texas, successfully lobbied the Texas State Board of Education to approve Mexican American studies as an elective in high school curricula. The Texas SBOE is a notoriously conservative body that has, in the past, eliminated whole chunks of civil rights history from Texas schoolbooks, so this curricular amendment is a major victory. We can also call it an environmental victory, in keeping with Parras's broad definition.

In my interview with Parras and Lopez, both note how they experienced the connections between literary advocacy and environmental justice organizing. Lopez emphasized that experiences with TEJAS helped pave the way for Librotraficante actions in practical ways. Parras notes how Nuestra Palabra helped him conceptualize the work TEJAS does in Houston and around the world. These two testimonials speak to how these two causes feed into and assist one another in focus, strategy, and message. They also demonstrate how Latina/o environmentalism is inextricably linked with social justice issues in the twenty-first century as much as, if not more than, in the past. Lopez said, "As far as connecting Librotraficante and Nuestra Palabra organizing with TEJAS organizing, I don't really see it as very different" (Parras and Lopez). Lopez also added that TEJAS's involvement with organizations such as the World Social Forum, and in particular their caravan tours throughout the United States on behalf of social justice, gave her and Parras the idea and willingness to organize the Librotraficante caravan: "Through that [Social Forum] organizing I could see how . . . it could be done . . . And it made planning the Librotraficante caravan in five weeks—to six cities, with workshops during the day, and banned book bashes in the evenings in five of those cities—it made it conceivable that we could do something like that" (Parras and Lopez). With this testimonial we can see how the environmental and social justice advocacy that Lopez and Parras took part in helped them to conceptualize the Librotraficante phenomenon.

In addition to the practical connections between the two groups, Parras adds that in either context it is important for them to add a fun, engaging, and even humorous dimension to their actions:

> We had the creativity to put something together that would resonate with our community and the public that was fun and wasn't the way you typically think of actions. It's hard to communicate that to folks who want to come and chain themselves to a truck or a fence. I think our communities respond to things that are a little more creative and involve some way of using our knowledge and bodies in a more fun way, with humor, and a critical pedagogy, but one that is also within the parameters of what the law allows us to do. We've seen too many of our fathers, uncles, brothers, and sisters get beat by the police and put away in

> prison or killed that we were taught "don't do that." What good are you to the larger community if you're dead, what good are you to the larger community if you're in prison? You have to be creative. (Parras and Lopez)

Parras details the importance of understanding your own community's sensibilities in order to design actions that will be effective at home and still communicate across cultural boundaries.

The conventional activist strategies that walk the fine line of legality do not always appeal to a community who already finds too many of its population behind bars. On one hand, as Diaz has expressed, the very idea of a "Librotraficante" is a sobering thought. The term should not make any sense in a more just world. On the other hand, the neologism is an ingenious riff on negative stereotypes circulated about Latinas/os that emphasizes our dedication to knowledge and learning, while also including the playfulness that appeals to our community.

In a more personal testimonial, Parras offered a perspective on how his involvement with Nuestra Palabra helped to facilitate his environmental justice advocacy with TEJAS. He volunteered with NP after graduating from college and found that the literary showcases introduced him to a world of Latina/o creativity that he had never known, not even during his university years. "There were a helluva lot more authors who looked like me and grew up with experiences similar to mine than I had ever imagined" (Lopez and Parras). Empowered by the proud voices surrounding him at those events—written and spoken by people who share his background—Parras found himself dedicated to helping Diaz fulfill his vision of creating a "Latino literary renaissance" in Houston in the early 2000s. Soon, he found himself volunteering on the weekly NP radio program at the local Pacifica station, KPFT. "I took what I learned from Nuestra Palabra and the radio show and began to apply that to what we were trying to do at TEJAS" (Lopez and Parras). At the same time, Parras's father was working for the Environmental Justice Legal Clinic at Texas Southern University and was becoming more involved in Houston's local cases, such as the city's decision to locate a new high school in a highly toxic and volatile neighborhood. TEJAS was born out of protesting this school's location, and Parras immediately saw how his NP organizing experience could be put to use toward TEJAS's goals.

These two groups, TEJAS and Nuestra Palabra, along with the successful Librotraficante actions, make up a campaign to fight the neoliberal assault on communities of color via attacks on our environments by contaminating them (toxicity) and eradicating them (decolonized knowledges). These groups refuse to let go of our environment's integrity, and they provide a useful example for other groups to follow through their practical experiences, culturally relevant practices, and personal journeys.

Some of the books destined to be taught in the new Mexican American studies classrooms in Texas that the Librotraficantes fought to create (and that were previously banned in Arizona after H.B. 2281 became law) were written by Luis Alberto Urrea and Ana Castillo. In the next section I discuss two of their works that engage the immigration debate and environmental issues on the U.S.-Mexico border. These two works testify to Chicana/o literature's ongoing interweaving of environmental and social justice issues.

THE DEVIL'S HIGHWAY AND *THE GUARDIANS*: DESERT MIGRANTS AND THE "AMERICAN DREAM"

The Devil's Highway and *The Guardians* center on U.S.-Mexico border migration and the growing tragedies in the desert regions of this political boundary. Since the 1990s, U.S. policy regarding border control (Operation Hold-the-Line 1993 and Operation Gatekeeper 1994) pushes migrants to attempt their crossings in the harshest and deadliest areas, such as the Sonoran Desert of the Arizona-Sonora boundary. The increase in migration in this area contributed to the penning of both S.B. 1070 and H.B. 2281. Two Chicana/o writers, Urrea and Castillo, take a nonfiction and a fiction approach to this debate. *The Devil's Highway* is Luis Alberto Urrea's 2004 nonfiction account of a group of twenty-six migrants who were abandoned by their "coyotes." Fourteen of them died in the Sonoran Desert. The book offers an analysis of the policies and economics on both sides of the border that lead to such tragedy. However, Urrea also provides a historical account of the many migrations the Sonoran Desert, and specifically the region known as the Devil's Highway, has seen across the years. Among them he includes the incursions of Spanish "explorers" and American "pioneers" to this region. He makes the point that, whether through circumstance or ignorance, human beings rarely belong in such a setting. Only the Tohono O'odham, who have cultivated survival there for generations, know how to survive, but even their homeland has been divided by the political border. The migrants from Mexico and Central America who now brave the harshest environments in pursuit of a better future for themselves and their children do not enjoy the celebrity that the pioneers and explorers now enjoy. Urrea puts their stories into context with those of past journeys. What makes those of the past who braved these deserts such heroes while those who do so today are branded criminals? *The Devil's Highway* draws together some of the harshest environments on the planet with some of the most desperate travelers to tell this contemporary story of human displacement and ineffective policies.

Undocumented travelers are labeled "illegal aliens." An illegal alien is not a pioneer or pilgrim, and is thus struck from the annals of great American migrations. Urrea points out how the Sonora-Arizona corridor has been a challenge for travelers for the entire span of recorded history, starting in the sixteenth century when Spain was colonizing the Sonoran Desert and when a European died in 1541. Urrea details the difficulty of the terrain, observing that the "plants are noxious and spiked" and the wildlife is "nocturnal . . . poisonous and alien" (Urrea 6). By 1850, one writer said that the Devil's Highway was "a vast graveyard of unknown dead" (12). On top of the climate, terrain, flora, and fauna, migrants in this region have to worry about competing coyote-gang turf. Urrea sums up his portrait of this region: "In the desert we are all illegal aliens" (120). He says this not to bolster the nativist label, but to change the context of the label. Humans all struggle to survive this climate. As literary critic Marta Caminero-Santangelo astutely observes about this sentence, Urrea "resituat[es] the 'we' so that the former 'they' is also included—a 'we' with boundaries flexible and fluid enough that 'illegal aliens' and U.S. citizens can inhabit it simultaneously" ("Narrating" 170). As humans with more in common than not, he argues, we need to see how our border and trade policies force our fellow humans to trek this terrain, all the while being despised for doing it.

Whatever happened to revering the migrant traveler in U.S. culture? For Urrea, it was Europeans who first orchestrated the northern migration from central Mexico to what is now the U.S.-Mexico border: "Immigration, the drive northward, is a white phenomenon. White Europeans conceived of and launched El Norte mania, just as White Europeans inhabiting the United States today bemoan it" (Urrea 8). The Europeans who moved north into Sonora were hailed as conquerors, those who landed on the east coast of North America were revered as pilgrims, and the Europeans who moved West were admired as pioneers. The current migration phenomenon is no longer "a white phenomenon," unless you credit the greedy trade and faulty immigration policies that push mestizos and indigenous peoples from their southern homelands to northern fields and factories of wage labor in today's Global North. Their lack of identification with whiteness calls this generation of migrants into question by the establishment in both Mexico and the United States, and puts these migrants into lethal situations.

The tragic death of fourteen migrants and the torturous survival of twelve others compelled Urrea to write this account. In the face of the great policy challenge that immigration presents, involving economic analyses and foreign relations, Urrea attempts to bring this problem down to the human scale. He takes readers into the minds of the walkers through the desert, and introduces us to their families back home in tropical Veracruz. He even humanizes the coyotes and the border patrol agents. If all the players in this drama are individuals with whom the reader can sympathize, then where is the villain? The analysis ultimately points to government trade deals that hobble small-scale farmers from making a living wage in Mexico and that force workers north for more lucrative but also degrading and dangerous labor in U.S. factories and fields. He also points to the fact that far from costing the United States for their use of public services, the United States makes a net profit from the taxes immigrants from Mexico pay. Urrea cites a 2003 study conducted by Thunderbird, the American Graduate School of International Management, which informs us that after accounting for Mexican immigrant contributions in the form of federal and sales taxes, minus the social services used by Mexican immigrants, the United States comes out with "a profit of $319 million" (Urrea 218). Urrea waits until the final pages of the book to present this information, slowly building his case against the criminalization of immigrants at the same time that he demystifies the Border Patrol and admires the courage of the migrants. Urrea exposes the nativist strategy as ineffective. The nativists reason that if they criminalize immigrants and mount even an environmental case against them, the United States will never hail their accomplishments and contributions. Urrea shows otherwise.

Ana Castillo takes up the immigration debate in a fictional context and moves east from Arizona to the greater El Paso and southeastern New Mexico region. She pens an early twenty-first-century realist novel, with touches of the magical, that focuses attention not only on immigration but also on the associated difficulties of drug and human trafficking, feminicide, gangs, environmental injustice, and poverty.

The Guardians, about a sister and son outside El Paso awaiting the undocumented crossing of their brother and father from Mexico, also asks why we criminalize the border crossers while many U.S. citizens benefit from the labor of undocumented workers and the black market of human trafficking and drug production. Using the arid setting of the Chihuahuan Desert and the Franklin Mountains, Castillo tells the story of a border in crisis—underfunded schools, youth in crisis, toxic environments, high murder rates, and difficulty making ends meet. Her narrative refuses to allow that such challenges are endemic to the U.S.-Mexico border, and instead it shows how these tragedies are linked to corrupt policies and prejudices. Together, Urrea's and Castillo's writings offer opportunities to engage with current issues of U.S.-Mexico border migration and consider the role of the desert both as a deadly environment and as a cultural and literary symbol.

To put it simply, *The Guardians* argues that we are in a crisis akin to New Testament perdition, requiring a savior in our midst. Instead of the Son of God, however, the twenty-first-century crisis gets a sixteen-year-old undocumented student named Gabriel, or Gabo. Gabo was born in Mexico and lives with his aunt Regina because his mother died crossing the border, and he is waiting for his father to return from a trip to Mexico. Gabo is a pious young Catholic man, which creates tension in his family because his father Rafa and his aunt Regina are suspicious of religion.

As the novel proceeds, Gabo becomes more and more like a Christ figure, testifying to his schoolmates and experiencing miracles such as the stigmata. At the same time, the problems mount in the rural community where Gabo and Regina live. Summer rains create sewage contamination in their friend Miguel's neighborhood, driving his family from their home, while Miguel's son's asthma worsens every day. Miguel's ex-wife Crucita gets kidnapped in Juarez, and everyone suspects feminicide. Even the sanctimonious Gabo gets into trouble with a local drug trafficking gang because he wants to find information about his missing father, and he cannot look to the local authorities for help. All together, these patterns of environmental injustice and violence on the border build in a way that mirrors New Testament proportions, and Gabo emerges as a potential savior. But the novel leaves readers wondering if any sacrifice will help a crisis of this proportion. As literary critic Theresa Delgadillo observes, "Gabriel and his aunt Regina harbor individual desires to combat the social disease around them through religious vocation and community improvement, respectively; their seeds of hope form important counterpoints to the world around them" ("Criticality of Latino/a Fiction" 616). This novel does not give in to a pessimistic view of the borderlands, but it gives readers an opportunity to observe the interconnected nature of the many challenges in this region at the same time that it presents empathetic characters who share their experiences in a rotating series of first-person narratives. Castillo employs this literary device to put readers at a variety of compelling vantage points. The dominant narrative becomes Regina's, and her community garden and other projects give the reader a sense of hope but also an idea of working together at the local scale to help address some of these challenges. The "Radio Nature" series on the nationally syndicated NPR radio program *LatinoUSA* shows communities across the country following this recipe in everyday life.

LATINOUSA'S "RADIO NATURE:" WHAT DO LATINA/O ENVIRONMENTALISTS SOUND LIKE?

As host María Hinojosa informs listeners, the "Radio Nature" segments of *LatinoUSA* are funded by a grant from the REI Foundation. *LatinoUSA* is a successful, nationally syndicated hour-long radio program that informs listeners every week about a variety of contemporary political, social, and cultural issues. The grant from REI is designed to fund *LatinoUSA* as it "documents the powerful and emotional bonds that diverse communities have with nature and is aimed at inspiring more outdoor enthusiasts and environmental stewards" (REI Foundation). Hinojosa describes the segments on the show as "stories of Latinos connecting to nature in powerful and transformative ways" (M. Hinojosa).

The segments have been airing since 2011 for a total of twenty-five "Radio Nature" stories. What first caught my attention about these stories is their mix of topics. I was immediately suspicious that the REI sponsorship might weigh these stories too heavily in the direction of depicting Latinas/os as just another group that can enjoy outdoor activities in the same way that white middle- and upper-class folks do. A good number of the stories do lean in this direction. I count fifteen out of the twenty-five with outdoor recreation as the central topic. These stories include the typical activities of hiking, fishing, and hunting, but also feature more adventurous activities such as spelunking in Puerto Rico, kayaking an urban New York City river, and summiting in the Himalayas. However, other topics show up, too, such as environmental injustice. My other concern was that these stories would focus on the Latina/o demographic as victims of environmental injustice. But this is not the case. The stories about environmental injustice, such as limited green space in urban areas and a proliferation of polluted neighborhoods, focus on the local communities' actions to improve their environments. These stories feature proactive leaders and hopeful futures. Finally, what most interested me was to hear stories that testified to Latina/o cultural knowledge and practices unique in the environmental context. My wish was granted with a number of stories that showcase traditional plant knowledge, agricultural practices passed down through the generations, and cultural elements such as "Danza Azteca with its constant references to agriculture and nature," as writer Raquel Z. Rivera puts it in the October 5, 2012, story titled "Planet New Mexico" (R. Rivera).

As a whole, so far, the stories come together to depict a diverse set of topics for Latinas/os where nature is concerned. For my taste, too many stories still focus on typical outdoor activities rather than proactive organizing to address environmental injustice or profiles of individuals or groups who put to use a wealth of traditional environmental knowledge. However, the segments shed light on the question that opened this book: why are Hispanics the demographic group most concerned with climate change? It turns out that our concern with climate change is rooted in our long-standing efforts to maintain a relationship between humans and nature that emphasizes dignity, respect, simplicity, and sustenance, and that transcends possession. Much work remains to uncover more of the ways of being and knowing that lend to this way of living in the world.

What do Latina/o environmentalists sound like? Maybe a better question for the twenty-first century is, "How can we better hear what Latinas/os are saying?" I expect that Latina/o decolonial environmentalisms will continue to develop and disperse throughout the twenty-first century and become a leading force in helping us all sustain a healthier life on this planet. The truth is that this book covers only a fraction of the work by Mexican Americans and Chicanas/os from which we can learn lessons for a better relationship with the natural environment. I am confident that this book is only one contribution to a long-lasting exchange and process of recovery and discovery. I also know that the way we have conceptualized environmentalism as separate from other social justice issues cannot last. Our environmental problems are a direct product of large-scale injustices that we can trace back to colonization and the darker side of modernity. The projects I include in this conclusion make this argument to differing degrees, but our success at building a better future relies on working to dismantle coloniality.

NOTES

INTRODUCTION

1. The Yale Project produces survey reports, peer-reviewed articles, and conference presentations. The specific Yale Project report that Maher and Leiserowitz discussed was *Race, Ethnicity, and the Public Response to Climate Change*. This report "provides an analysis of the global warming and energy policy preferences of both the overall American public and of particular racial and ethnic groups" (Leiserowitz and Akerlof 4).
2. My thanks go to Yolanda Flores Niemann, who brought to my attention this particular episode of *Real Time*.
3. Among the most notable exceptions are Devon Peña's two books, *Mexican Americans and the Environment: Tierra y Vida* and *Chicano Culture, Ecology, Politics: Subversive Kin*, as well as Laura Pulido's *Environmentalism and Economic Justice: Two Chicano Struggles in the Southwest*. Scholars Kamala Platt, María Herrera-Sobek, Patrick D. Murphy, Tom Lynch, Carmen Flys-Junquera, Alex Hunt, and Christa Grewe-Volp have all published journal articles or book chapters about Chicana/o literature and the environment.
4. The six polls include a Natural Resources Defense Council poll conducted by Latino Decisions; a poll collectively conducted by the *New York Times*, Stanford University, and Resources for the Future; a poll commissioned by the California League for Conservation Voters; a poll conducted by the

Sierra Club in conjunction with the National Council of La Raza; a poll by American Climate Values and EcoAmerica; and a poll conducted by Green for All.

5. Two terms are used today to describe the Spanish-speaking population in the United States: Hispanic and Latina/o. I will not use the term "Hispanic" because of its history of invoking European heritage to the detrimental exclusion of indigenous and mestiza/o identities. If referring to the larger Spanish-speaking population of the United States, I will use the term "Latina/o." This book, however, takes up the writings of a particular group within the Latina/o population: Mexican Americans and Chicanas/os. The distinction between Mexican American and Chicana/o in this work relates to chronology: from 1848 until the Civil Rights Movement, this ethnic group was known as Mexican American. During the Civil Rights Movement, Mexican Americans took the name Chicanas/os to identify themselves more closely with their indigenous background and to emphasize a progressive political agenda. In focusing on Mexican American and Chicana/o writing, I do not mean to suggest that they are the Latina/o population most concerned with the natural environment. My emphasis emerges from my expertise in this body of work, and I hope that scholars with expertise in the literatures of other Latina/o populations will also develop readings of the environment in those writings.
6. In an excellent chapter on "Green Aztlán: Environmentalism and the Chicano/a Visual Arts" in his book *In the Spirit of a New People: The Cultural Politics of the Chicano Movement*, Randy Ontiveros also observes that "[e]nvironmental awareness among racial minorities is not new" (87).
7. Readers may be familiar with the phrase "the good life" from the writings of Helen and Scott Nearing, who recount their experience of leaving New York City in 1932 to live a life of simplicity in rural Vermont and Maine. They are hailed as the "great-grandparents of the back-to-the-land movement." Their book *Living the Good Life* was published in 1970. Although the Nearings and Cabeza de Baca share the values of simplicity and sustenance, I contrast my use of "goodlife" with theirs on the basis that Cabeza de Baca gives a name in 1949 to a set of practices and beliefs that have been ongoing for Mexican Americans ever since at least the late nineteenth century. The Nearings, on the other hand, introduce into U.S. culture a set of practices and values meant to challenge the predominant way of life for mainstream America in 1970.
8. See the volume *Aldo Leopold's Southwest*, edited by David E. Brown and Neil B. Carmony, as well as the 2009 symposium, "A Cultural Conversation: Aldo

Leopold, the Southwest, and the Evolution of a Land Ethic," at the National Hispanic Cultural Center in Albuquerque.

9. One standout work in this regard is Sarah Jacquette Ray's *The Ecological Other: Environmental Exclusion in American Culture.*
10. See Morton for insightful and accessible discussions of this point. His book *Ecology Without Nature* explores "what we mean by the word *environment* itself" by tracing its genealogy back to the Romantics' writings on nature, and he argues that the Western notion of nature itself gets in the way of sustaining an ecology favorable for human habitation (3).
11. To cite just one example: a University of California–Irvine research team has shown how there is presently a "disproportionate proximity to toxic waste among Latinos" in Southern California (Hipp and Lakon 2010, 681).
12. See Lynch 258–59 for an exposition of acequia culture in Baca.
13. As Kamala Platt notes, Castillo's protagonist "slowly loses her faith in the benign-ness of the system, which she embraced because it supported her (literal) buying into a related system—mass consumer culture. Thus Castillo underscores the importance of personal responsibility for lifestyle in reducing the global vulnerability of environment destruction" (87). An "ecological footprint" consists of a population's impact on natural resources. The United States consumes far more resources than locally available to its population, making its ecological footprint disproportionately large.
14. See Wald, Shea, and Grewe-Volpp for excellent analyses of farm labor in Viramontes's brilliant text.
15. Current and emerging scholarship in comparative indigeneities takes issue with the exclusive references to Aztec/Mixtec ancestry that erases the multitude of indigenous groups in the Americas. This scholarship also objects to the invocation of indigenous mythology as if it were all but extinct, while so many indigenous groups and their beliefs exist and actively nurture their cultural practices today. See Alberto for more on this topic.
16. Nine out of the fourteen contributors in the 1999 *PMLA* "Forum on Literatures of the Environment" in one way or another lamented the lack of diversity within the ranks of ecocriticism. They call attention to the Anglocentricity of ecocriticism, in reference both to the makeup of the ranks of ecocritics themselves as well as the writers whom ecocritics choose to study. Indeed, in her letter, Elizabeth Dodd quotes field-founder Cheryll Glotfelty's observation that "[e]cocriticism has been predominantly a white movement" (Parra 1094). Two of these nine scholars, Terrell Dixon and Andrea Parra,

cite specific Chicana and Chicano environmental writings. Dixon discusses the ways that the writings of South Texas borderlanders Arturo Longoria and Gloria Anzaldúa "illustrate both the environmental destruction and the cultural origins of that destruction in national borderlands created to enforce ethnic, economic, and class divisions" and explains ways that "these two texts also suggest the possibilities for effective restoration that can come with natural, instead of national, boundaries" (Dixon 1094). Parra emphasizes what she calls ecofeminist Chicana writings, especially the works of Ana Castillo, Cherríe Moraga, and Gloria Anzaldúa. She also aptly argues that the "[r]edefinition of *the environment* to include both urban and rural landscapes will allow for the critical study of not only the desert flowers of Pat Mora but also the freeways of Lorna Dee Cervantes and the suburban dumps and strip malls evoked by Junot Diaz's fiction" (Parra 1100). Both Dixon and Parra make a clear case for the richness of contemporary Chicana/o writing's potential for ecocritical focus, and Parra's mention of Junot Diaz shows that Dominicans and other Latino groups have a lot to say as well.

17. One notable exception is Greta Gaard's essay "Toward a Queer Ecofeminism," in which she shows how the "rhetoric and institution of Christianity, coupled with the imperialist drives of militarist nation-states, have been used for nearly two thousand years to portray heterosexuality, sexism, racism, classism, and the oppression of the natural world as divinely ordained" (31).
18. See Liptak for details on how the Supreme Court struck down the portion of the VRA that required nine states, mostly in the South, to seek federal approval before changing their election laws.
19. Miguel Angel Porta-Gandara, Eduardo Rubio, and Jose Luis Fernandez, "Economic Feasibility of Passive Ambient Comfort in Baja California Dwellings," *Building and Environment* 37 (2002): 993. "The results indicate that, in cases as those analyzed, the use of vernacular passive techniques [adobe homes] is more comfortable and economic than present light buildings by a very wide margin."

CHAPTER 1

1. See Foley 31–33 and Griswold del Castillo.
2. The 1996 editors of *Caballero* include this argument in an epilogue written by María Cotera, "Hombres Necios: A Critical Epilogue," where she says the

novel "presents an oppositional response both to dominant patriarchal culture as a whole and to elements of that culture in traditional Chicano texts" (339).

3. Literary critic Randy Ontiveros, building on Jake Kosek's work, notes that Muir's environmental politics enacted his own prejudices: "Muir argued for white guardianship of the land, championing the U.S. Army's occupation of Yosemite as a deterrent against Indian degradation, and naively ignoring the long-term damage that scientific control and mass tourism would do to the continent's majestic wilderness" (91).

CHAPTER 2

1. See John González's *Border Rennaisance* for readings of this period as radical.
2. Walter Mignolo discussed this difficulty in his lectures at Southern Methodist University and the University of North Texas in spring 2014.
3. See Charlotte Whaley for a comprehensive biography of Nina Otero Warren.
4. Rebolledo offers analyses of Otero Warren in her 1989 essay "Las Escritoras," as well as in her 1995 monograph *Women Singing in the Snow.*
5. Her subsequent novel, *Dew on the Thorn*, tells the story of one ranch, its owners, and its workers—known as "patrones" and "peones," respectively, in the semifeudal system of the early nineteenth century. Her coauthored novel with Eve Raleigh, *Caballero*, also tells the story of a ranch family, but it is set in the tumultuous time of the U.S.-Mexican War (1846–1848), and as a result shows the impact that the war had on a family and a traditional community. See chapter 1. Unfortunately, neither novel ever found a publisher during her lifetime. However, both novels were published posthumously, in the 1990s, thanks to the recovery work of José Limón and María Cotera. She did, however see some of her work published in the annual collections distributed by the Texas Folklore Society, for whom she was elected president for the years 1930 through 1932.
6. See Cotera, Limón, and Reyna.
7. *Dew on the Thorn* remained unpublished until it was recovered by José Limón and published by Arte Público Press in Houston, Texas.

CHAPTER 3

1. See Nabokov.
2. See U.S. General Accounting Office (GAO) report.
3. See Tijerina.
4. See GAO report.
5. In 2009, Ken Burns and PBS released a series of documentary films titled *The National Parks: America's Best Idea.*
6. See Theresa Delgadillo's *Spiritual Mestizaje: Religion, Gender, Race, and Nation in Contemporary Chicana Narrative* for a recent authoritative study on spirituality and its critique of oppression in Chicana writing.

CHAPTER 4

1. See Mendoza.
2. The text I quote in this chapter is from Evangelina Vigil Piñón's translation for Arte Público Press's 1995 bilingual edition of this novel, which I use to teach and is useful for bilingual students.
3. In his essay, Carlos Gallego recommends the lens of cognitive mapping over what he calls the "tired themes" of community consciousness and Chicana/o solidarity.
4. See Fiskio for an insightful essay on the fetishization of place-based environmentalism and what can be learned from factoring migrant experiences into environmental values and practices.
5. I quote the lyrics throughout the essay. Here they are in their entirety:

   "Something in the Rain" by Tish Hinojosa
   Mom and Dad have worked the fields
   I don't know how many years
   I'm just a boy but I know how
   And go to school when work is slow
   We have seen our country's roads
   Bakersfield to Illinois
   And when troubles come our way,
   Oh yeah, I've seen my daddy pray

There's something wrong with little sister
I hear her crying by my side
Mama's shaking as she holds her
We try to hold her through the night
And Mom says close your eyes, *mi'jito*
Dream of someplace far from here
Like the pictures in your school books
Someday you can take us there

There must be something in the rain
I'm not sure just what that means
Abuelita talks of sins of man
Of dust that's in our hands
There must be something in the rain,
Well, what else could cause this pain?
Those airplanes cure the plants so things can grow
Oh no, it must be something in the rain

Little sister's gone away
Mama's workin' long again
And me, I think I understand
About our life, about our land
Well, talkers talk and dreamers dream
I will find a place between
I'm afraid but I believe
That we can change these hurting fields

Cause there's something in the rain
But there's more here in our hands
'Buelita's right about the sins of man
Whose profits rape the land
And the rains are pouring down
From the growers to the towns
And until we break the killing chains
There's something in the rain

6. For especially strong essays on this novel, see Moya "Reading as a Realist," Shea, and Grewe-Volpp.

CHAPTER 5

1. Randy Ontiveros argues that the manifesto Alurista wrote for the 1969 Denver Youth Conference, El Plan Espiritual de Aztlán, reveals that environmental values fundamentally shape the then-emerging Chicano nationalist consciousness: "Alurista envisions an alternative economy built on the recognition of a shared moral obligation to 'our lands' and to 'those who plant the seeds, water the fields, and gather the crops.' He calls this alternative Aztlán, and he imagines it a place governed not by profit margins and efficiencies, but by 'the process of love and brotherhood.'" (Ontiveros 97)

   I agree with Ontiveros's reading here, and I have made a similar argument elsewhere (Marcone and Ybarra). However, the Chicana/o cultural nationalism that becomes most visible at this time period and through today has more to do with appropriating land than it does with transcending possession. This chapter brings attention back to Chicana/o cultural nationalism that better resonates with the values found in the Plan Espiritual de Aztlán.
2. For more on this topic see Stanley Brandes, "Sugar, Colonialism, and Death," as well as "Iconography in Mexico's Day of the Dead."

WORKS CITED

Abbey, Edward. *Desert Solitaire*. 1968. New York: Ballantine, 1972.

Adamson, Joni, Mei Mei Evans, and Rachel Stein, eds. *The Environmental Justice Reader: Politics, Poetics, and Pedagogy*. Tucson: University of Arizona Press, 2002.

Alberto, Lourdes. "Topographies of Indigenism: Mexico, Decolonial Indigenism, and the Chicana Transnational Subject in Ana Castillo's *Mixquiahuala Letters*." *Comparative Indigeneities of the Américas: Toward a Hemispheric Approach*. Eds. M. Bianet Castellanos, Lourdes Gutiérrez Nájera, and Arturo J. Aldama. Tucson: University of Arizona Press, 2012. 38–52.

Alemán, Jesse. "'Thank God, Lolita Is Away from Those Horrid Savages': The Politics of Whiteness in *Who Would Have Thought It?*" *María Amparo Ruiz de Burton: Critical and Pedagogical Perspectives*. Eds. Amelia María de la Luz Montes and Anne Elizabeth Goldman. Lincoln: University of Nebraska Press, 2004. 95–111.

Anaya, Rudolfo. *Bless Me, Ultima*. Berkeley: Quinto Sol, 1972.

Anzaldúa, Gloria. *Borderlands/La Frontera: The New Mestiza*. San Francisco: Aunt Lute Books, 1987.

Aranda, José F. *When We Arrive: A New Literary History of Mexican America*. Tucson: University of Arizona Press, 2003.

Arizona. Legislature. House. House Bill 2281. *An Act Amending title 15, chapter 1, article 1, Arizona Revised Statutes, by adding sections 15-111 and 15-112; amending section 15-843, Arizona Revised Statutes; relating to school curriculum*. State Law. 1–4. (49th Legislature 2010). *Arizona State Legislature*. Web. 10 Aug. 2012.

Arizona. Legislature. Assembly. Senate Bill 1070. *An Act Amending title 11, chapter 7, Arizona Revised Statutes, by adding article 8; amending title 13, chapter 15, Arizona Revised Statutes, by adding section 13-1509; amending section 13-2319, Arizona Revised Statutes; amending title 13, chapter 29, Arizona Revised Statutes, by adding sections 13-2928 and 13-2929; amending sections 23-212, 23-212.01, 23-214 and 28-3511, Arizona Revised Statutes; amending title 41, chapter 12, article 2, Arizona Revised Statutes, by adding section 41-1724; relating to unlawfully present aliens*. State Law. 1–16. (49th Legislature 2010). *Arizona State Legislature*. Amending Title 11, Chapter 7. Web. 10 Aug. 2012.

Arnold, Jean, Lawrence Buell, Michael P. Cohen, Terrell Dixon, Andrea Parra, et al. "Forum on Literatures of the Environment." *PMLA: Publications of the Modern Language Association of America* 114.5 (1999): 1089–104.

Arrizón, Alicia. "Mythical Performativity: Relocating Aztlan in Chicana Feminist Cultural Productions." *Theatre Journal* 52.1 (2000): 23–49.

Baca, Jimmy Santiago. *Black Mesa Poems*. New York: New Directions, 1989.

Bierhorst, John. *The Hungry Woman: Myths and Legends of the Aztecs*. New York: Harper Collins, 1993.

Blackwell, Maylei. *¡Chicana Power!: Contested Histories of Gender and Feminism in the Chicano Movement*. Austin: University of Texas Press, 2011.

Brady, Mary Pat. *Extinct Lands, Temporal Geographies: Chicana Literature and the Urgency of Space*. Durham, NC: Duke University Press, 2002.

Brandes, Stanley. "Iconography in Mexico's Day of the Dead: Origins and Meanings." *Ethnohistory* 45.2 (1998): 181–218.

———. "Sugar, Colonialism, and Death: On the Origins of Mexico's Day of the Dead." *Comparative Studies in Society and History* 39.2 (1997): 270–99.

Brown, David E., and Neil B. Carmony, eds. *Aldo Leopold's Southwest*. Albuquerque: University of New Mexico Press, 1990.

Buell, Lawrence. "The Ecocritical Insurgency." *New Literary History* 30.3 (1999): 699–712.

———, Ursula Heise, and Karen Thornber. "Literature and Environment." *Annual Review of Environment and Resources* 36 (2011): 417–40.

Burford, Arianne. "Cartographies of a Violent Landscape: Viramontes' and Moraga's Remapping of Feminisms in *Under the Feet of Jesus* and *Heroes and Saints*." *Genders* 47 (2008). Web.

Cabeza de Baca, Fabiola. *The Good Life: New Mexico Traditions and Food.* 1949. Albuquerque: Museum of New Mexico Press, 2005.

———. *We Fed Them Cactus.* 1954. Albuquerque: University of New Mexico Press, 1994.

Cafaro, Philip. *How Many Is Too Many?: The Progressive Argument for Reducing Immigration into the United States.* Chicago: University of Chicago Press, 2015.

Cafaro, Philip, and Winthrop Staples III. "The Environmental Argument for Reducing Immigration to the United States." *Environmental Ethics* 31.1 (2009): 5–30.

Callicott, J. Baird, and Michael P. Nelson, eds. *American Indian Environmental Ethics: An Ojibwa Case Study.* Upper Saddle River, NJ: Prentice Hall, 2004.

Caminero-Santangelo, Marta. "Narrating the Non-nation: Literary Journalism and 'Illegal' Border Crossings." *Arizona Quarterly: A Journal of American Literature, Culture, and Theory* 68.3 (Autumn 2012): 157–76.

———. "The Lost Ones: Post-Gatekeeper Border Fiction and the Construction of Cultural Trauma." *Latino Studies* 8.3 (2010): 304–27.

———. "'Jasón's Indian': Mexican Americans and the Denial of Indigenous Ethnicity in Anaya's *Bless Me, Ultima.*" *Critique: Studies in Contemporary Fiction* 45.2 (2004): 115–28.

Campa, A. L. "*Old Spain in Our Southwest* by Nina Otero." *The Journal of American Folklore* 51.200 (1938): 208–9.

Carson, Rachel. *Silent Spring.* 1962. New York: Houghton Mifflin, 2002.

Casas, María R. *Married to a Daughter of the Land: Spanish-Mexican Women and Interethnic Marriage in California, 1820–1880.* Reno: University of Nevada Press, 2007.

Castillo, Ana. *So Far from God.* New York: Plume, 1994.

———. *The Guardians.* New York: Random House, 2007.

Chambers, William T. "Lower Rio Grande Valley of Texas." *Economic Geography* 6.4 (1930): 364–73.

Chávez, Denise. "Crossing Bitter Creek: Meditations on the Colorado River." *Writing Down the River: Into the Heart of the Grand Canyon.* Ed. Kathleen Jo Ryan. Flagstaff, AZ: Northland, 1998.

Chávez, John R. *The Lost Land: The Chicano Image of the Southwest.* Albuquerque: University of New Mexico Press, 1984.

Cohen, Michael P. "Blues in the Green: Ecocriticism Under Critique." *Environmental History* 9.1 (2004): 9–36.

Cotera, María Eugenia. "Refiguring 'The American Congo': Jovita González, John Gregory Bourke, and the Battle over Ethno-historical Representations of the Texas Mexican Border." *Western American Literature* 35.1 (2000): 75–94.

———. "Hombres Necios: A Critical Epilogue." *Caballero: A Historical Novel*. Eds. José E. Limón and María Cotera. College Station: Texas A&M University Press, 1996.

———. "Jovita González Mireles: A Sense of History and Homeland." *Latina Legacies: Identity, Biography, and Community*. Eds. Vicki L. Ruiz and Virginia Sánchez Korrol. Oxford: Oxford University Press, 2005. 158–74.

Coulter, Ann. *¡Adios, America!: The Left's Plan to Turn Our Country into a Third World Hellhole*. Washington, DC: Regnery Publishing, 2015.

Cronon, William. "The Trouble with Wilderness, or, Getting Back to the Wrong Nature." *The Great New Wilderness Debate*. 1995. Eds. J. Baird Callicott and Michael Nelson. Athens: University of Georgia Press, 1998. 471–99.

A Cultural Conversation: Aldo Leopold, the Southwest, and the Evolution of a Land Ethic for the Future (Conference). National Hispanic Cultural Center, Albuquerque, New Mexico. 13–14 Feb. 2009.

Davenport, Coral. "Climate is Big Issue for Hispanics, and Personal." *New York Times on the Web* 9 Feb. 2015.

Davis, Mike. "Los Angeles After the Storm: Dialectic of Ordinary Disaster." *Antipode* 27.3 (1995): 221–41.

DeLoughrey, E. M., R. K. Gosson, and G. B. Handley, eds. *Caribbean Literature and the Environment: Between Nature and Culture*. Charlottesville: University of Virginia Press, 2005.

Delgadillo, Theresa. *Spiritual Mestizaje: Religion, Gender, Race, and Nation in Contemporary Chicana Narrative*. Durham, NC: Duke University Press, 2011.

———. "The Criticality of Latino/a Fiction in the Twenty-First Century." *American Literary History* 23.3 (Fall 2011): 600–624.

Escobar, Arturo. "Sustainability: Design for the Pluriverse." *Development* 54.2 (2011): 137–40.

Fish, Stanley. "Arizona: The Gift that Keeps On Giving." Editorial. *New York Times on the Web* 17 May 2010.

Fiskio, Janet. "Unsettling Ecocriticism: Rethinking Agrarianism, Place, and Citizenship." *American Literature* 84.2 (2012): 301–25.

Foley, Neil. *Mexicans in the Making of America*. Cambridge, MA: Harvard University Press, 2014.

Foscue, Edwin J. "Land Utilization in the Lower Rio Grande Valley of Texas." *Economic Geography* 8.1 (1932): 1–11.

Fox, Stephen. *John Muir and His Legacy: The American Conservation Movement.* Boston: Little, Brown, 1981.

Freinkel, Susan. "Warning Signs." *The Nation* 31 March 2014: 12–22.

Gaard, Greta. "Toward a Queer Ecofeminism." *New Perspectives on Environmental Justice: Gender, Sexuality, and Activism.* New Brunswick, NJ: Rutgers University Press, 2004. 21–44.

Gallego, Carlos. "Topographies of Resistance: Cognitive Mapping in Chicana/o Migrant Literature." *Arizona Quarterly: A Journal of American Literature, Culture, and Theory* 70.2 (2014): 21–53.

Garrard, Greg. *Ecocriticism.* London: Routledge, 2011.

Gaspar de Alba, Alicia. *Chicano Art Inside/Outside the Master's House: Cultural Politics and the CARA Exhibition.* Austin: University of Texas Press, 1998.

Gifford, Terry. "Recent Critiques of Ecocriticism." *New Formations* 64 (2008): 15–24.

Gómez, Laura E. *Manifest Destinies: The Making of the Mexican American Race.* New York: New York University Press, 2007.

Gonzales, Rodolfo "Corky." "I Am Joaquín (Yo Soy Joaquín)." *Herencia: The Anthology of Hispanic Literature of the United States.* Ed. Nicolás Kanellos. Houston: Arte Público Press, 2002.

———. *Message to Aztlán: Selected Writings.* 1965. Houston: Arte Público Press, 2001.

González, John M. *Border Renaissance: The Texas Centennial and the Emergence of Mexican American Literature.* Austin: University of Texas Press, 2010.

———. "The Whiteness of the Blush: The Cultural Politics of Racial Formation in *The Squatter and the Don.*" *María Amparo Ruiz de Burton: Critical and Pedagogical Perspectives.* Eds. Amelia María de la Luz Montes and Anne Elizabeth Goldman. Lincoln: University of Nebraska Press, 2004.

González, Jovita. *Dew on the Thorn.* Ed. José E. Limón. Houston: Arte Público Press, 1997.

———. "Folklore of the Texas-Mexican Vaquero." *Texas and Southwestern Lore.* Ed. J. Frank Dobie. Austin: Texas Folklore Society, 1927. 7–22.

———. *The Woman Who Lost Her Soul and Other Stories.* Ed. Sergio Reyna. Houston: Arte Público Press, 2000.

González, Jovita, and Eve Raleigh. *Caballero: A Historical Novel.* Ed. José Limón and María Cotera. College Station: Texas A&M Press, 1996.

Green for All. *Climate Change and Communities of Color*. 2014. 20 March 2015.

Greenberg, Linda Margarita. "Learning from the Dead: Wounds, Women, and Activism in Cherríe Moraga's *Heroes and Saints*." *MELUS: Multi-Ethnic Literature of the United States* 34.1 (2009): 163–84.

Grewe-Volpp, Christa. "'The Oil Was Made from Their Bones': Environmental (In)Justice in Helena María Viramontes's *Under the Feet of Jesus*." *ISLE: Interdisciplinary Studies in Literature and Environment* 12.1 (2005): 61–78.

Griswold del Castillo, Richard. *The Treaty of Guadalupe Hidalgo: A Legacy of Conflict*. Norman: University of Oklahoma Press, 1990.

Guha, Ramachandra. "Radical American Environmentalism and Wilderness Preservation: A Third World Critique." *Environmental Ethics* 11 (1989): 71–83.

Hames-Garcia, Michael. "How to Tell a Mestizo from an Enchirito®: Colonialism and National Culture in the Borderlands." *Diacritics* 30.4 (2000): 102–22.

Hardt, Michael, and Antonio Negri. "Sovereignty of the Nation-State." *Empire*. Cambridge, MA: Harvard University Press, 2000.

Herrera-Sobek, María. "The Nature of Chicana Literature: Feminist Ecological Literary Criticism and Chicana Writers." *Revista Canaria de Estudios Ingleses* 37 (1998): 89–100.

———. "Writing the Toxic Environment: Ecocriticism and the Chicana Literary Imagination." *A Contested West: New Readings of Place in Western American Literature*. Eds. Martin Simonson, David Rio, and Amaia Ibarran. London: BM Portal Editions, 2013. 173–87.

Hinojosa, María, host. *LatinoUSA*. National Public Radio. KERA, Dallas. 5 Oct. 2012.

Hinojosa, Tish. "Something in the Rain." *Culture Swing*. Rounder Records, 1992.

Hipp, John R., and Cynthia M. Lakon. "Social Disparities in Health: Disproportionate Toxicity Proximity in Minority Communities over a Decade." *Health & Place* 16 (2010): 674–83.

Holmes, Seth M. *Fresh Fruit, Broken Bodies: Migrant Farmworkers in the United States*. Berkeley: University of California Press, 2013.

Hunt, Alex. "In Search of Anaya's Carp: Mapping Ecological Consciousness and Chicano Myth." *ISLE* 12.2 (2005): 179–206.

Jackson, Helen Hunt. *Ramona, a Story*. Boston: Robert Brothers, 1884.

Joyce, James. *A Portrait of the Artist as a Young Man*. 1916. New York: W. W. Norton, 2007.

Karttunen, Frances. "Zenzontle." *An Analytical Dictionary of Nahuatl*. Austin: University of Texas Press, 1983. 32.

Keller, Robert H., and Michael F. Turek. *American Indians & National Parks.* Tucson: University of Arizona Press, 1998.

Kosek, Jake. *Understories: The Political Life of Forests in Northern New Mexico.* Durham, NC: Duke University Press, 2006.

Latino Decisions. *Natural Resources Defense Council Latino Decisions Survey—January 2014.* 15 Jan. 2015.

Leiserowitz, Anthony, and Karen Akerlof. *Race, Ethnicity, and Public Response to Climate Change.* New Haven, CT: Yale Project on Climate Change, 2010.

Leopold, Aldo. *A Sand County Almanac and Sketches Here and There.* New York: Oxford University Press, 1949.

Limón, José E. *Dancing with the Devil: Society and Cultural Poetics in Mexican-American South Texas.* Madison: University of Wisconsin Press, 1994.

———. Introduction. *Dew on the Thorn.* By Jovita González. Houston: Arte Público Press, 1997. xv–xxviii.

Liptak, Adam. "Supreme Court Invalidates Key Part of Voting Rights Act." *New York Times on the Web* 25 June 2013.

Longoria, Arturo. *Adios to the Brushlands.* College Station: Texas A&M University Press, 1997.

López, Marissa K. *Chicano Nations: The Hemispheric Origins of Mexican American Literature.* New York: New York University Press, 2011.

Lugones, María. *Pilgrimages/Peregrinajes: Theorizing Coalition Against Multiple Oppressions.* New York: Rowman and Littlefield, 2003.

Lynch, Thomas P. "Toward a Symbiosis of Ecology and Justice: Water and Land Conflicts in Frank Waters, John Nichols, and Jimmy Santiago Baca." *The Environmental Justice Reader: Politics, Poetics, and Pedagogy.* Eds. Joni Adamson, Mei Mei Evans, and Rachel Stein. Tucson: University of Arizona Press, 2002. 247–64.

Maher, Bill. *Real Time with Bill Maher.* 19 June 2013. Home Box Office. Television.

———. *Real Time with Bill Maher.* 28 June 2015. Home Box Office. Television.

Maldonado-Torres, Nelson. "On the Coloniality of Being." *Cultural Studies* 21.2–3 (2007): 240–70.

Marcone, Jorge. "A Painful Pastoral: Migration and Ecology in Chicana/o Literature." *Pacific Coast Philology* 46.2 (2011): 194–209.

Marcone, Jorge, and Priscilla Solis Ybarra. "Inhabiting and Unearthing: Chicana/o and Mexican Environmental Writing." *Teaching North American Environmental Literature.* Eds. Laird Christensen, Mark C. Long, and Fred Waage. New York: Modern Language Association Publications, 2008. 93–111.

Martin, Patricia Preciado. "Earth to Earth." *Days of Plenty, Days of Want*. Tucson: University of Arizona Press, 1999.

———. *Songs My Mother Sang to Me: An Oral History of Mexican American Women*. Tucson: University of Arizona Press, 1992.

Martín-Alcoff, Linda. "In Arizona, Censoring Questions About Race." Editorial. *New York Times on the Web* 1 Apr. 2012.

Martinez, Elizabeth. "El gran robo de nuestra tierra." *El Grito del Norte* 7 Dec. 1970. 12–13. Hispanic American Newspapers Database.

McEntyre, Mary Chandler. "Sickness in the System: The Health Costs of the Harvest." *Journal of Medical Humanities* 28.2 (2007): 97–104.

Menchaca, Martha. *Naturalizing Mexican Immigrants: A Texas History*. Austin: University of Texas Press, 2011.

Mendoza, Louis Gerard. *Historia: The Literary Making of Chicana and Chicano History*. College Station: Texas A&M University Press, 2001.

Merchant, Carolyn. *Ecological Revolutions: Nature, Gender, and Science in New England*. Chapel Hill: University of North Carolina Press, 1989.

Mignolo, Walter. "The Geopolitics of Knowledge and the Colonial Difference." *Coloniality at Large: Latin America and the Postcolonial Debate*. Eds. Mabel Moraña, Enrique Dussel, and Carlos A. Jáuregui. Durham, NC: Duke University Press, 2008.

———. *The Darker Side of Western Modernity: Global Futures, Decolonial Options*. Durham, NC: Duke University Press, 2011.

———. "Enacting the Archives, Discentering the Muses: Museums and Biennials in the Colonial Horizon of Modernity." Lecture. Southern Methodist University, Dallas, Texas. 30 Jan. 2014.

———. "The Closing of Modernity: Growth, Development, Death." Lecture. University of North Texas, Denton, Texas. 17 Apr. 2014.

Momentus. *American Climate Values 2014: Psychographic and Demographic Insights*. March 2014. http://ecoamerica.org/wp-content/uploads/2014/04/eA_American_Climate_Values_2014_Psychographic_and_Demographic_Insights.pdf. 16 Jan. 2015.

Montejano, David. *Anglos and Mexicans in the Making of Texas, 1836–1986*. Austin: University of Texas Press, 1987.

Moraga, Cherríe. *Heroes and Saints & Other Plays*. Albuquerque, NM: West End Press, 1994.

———. *The Hungry Woman: A Mexican Medea*. Albuquerque, NM: West End Press, 2001.

———. "Indígena as Scribe: The (W)Rite to Remember." *A Xicana Codex of Changing Consciousness: Writings, 2000–2010*. Durham, NC: Duke University Press, 2011.

———. *The Last Generation: Prose & Poetry*. Boston: South End Press, 1993.

———. *Loving in the War Years*. Boston: South End Press, 1983.

Moraga, Cherríe, and Gloria Anzaldúa, eds. *This Bridge Called My Back: Writings by Radical Women of Color*. New York: Kitchen Table Press, 1983.

Mortimer-Sandilands, Catriona. "Melancholy Natures, Queer Ecologies." *Queer Ecologies: Sex, Nature, Politics, Desire*. Bloomington: Indiana University Press, 2010. 331–58.

Morton, Timothy. *Ecology Without Nature: Rethinking Environmental Aesthetics*. Cambridge, MA: Harvard University Press, 2009.

Moya, Paula M. L. "Who We Are and from Where We Speak." *Transmodernity: Journal of Peripheral Cultural Production of the Luso-Hispanic World*. 1.2 (2011): 79–94.

———. "Reading as a Realist: Expanded Literacy in Helena María Viramontes's *Under the Feet of Jesus*." *Learning from Experience: Minority Identities, Multicultural Struggles*. Berkeley: University of California Press, 2002. 175–214.

Muir, John. *A Thousand Mile Walk to the Gulf*. New York: Houghton Mifflin, 1916.

———. *Steep Hills*. Boston: Houghton Mifflin, 1918.

———. Letter from John Muir. *The Letters of Ralph Waldo Emerson*. Ed. Ralph L. Rusk. New York: Columbia University Press, 1939. 155.

Mullen, Harryette. "'When He Is Least Himself': *Double Consciousness* in African American Poetry." *African American Review* 41.2 (2007): 277–82.

Muñoz, José Esteban. *Cruising Utopia: The Then and There of Queer Futurity*. New York: New York University Press, 2009.

Murphy, Andrew R. "Environmentalism, Antimodernism, and the Recurrent Rhetoric of Decline." *Environmental Ethics* 25 (Spring 2003): 79–98.

Murphy, Patrick D. *Farther Afield in the Study of Nature-Oriented Literature*. Charlottesville: University Press of Virginia, 2000.

Myers Research. *Nationwide Study of Hispanic Voters and the Environment*. 2012. Sierra Club and National Council of La Raza. 8 May 2015.

Nabokov, Peter. *Tijerina and the Courthouse Raid*. Albuquerque: University of New Mexico Press, 1969.

The National Parks: America's Best Idea. Dir. Ken Burns. PBS. 2009. Television.

Nearing, Helen, and Scott Nearing. *Living the Good Life: How to Live Sanely and Simply in a Troubled World*. New York: Schocken Books, 1970.

Nieto-Phillips, John M. *The Language of Blood: The Making of Spanish-American Identity in New Mexico, 1880s–1930s*. Albuquerque: University of New Mexico Press, 2004.

Olguín, B. V. "Caballeros and Indians: Mexican American Whiteness, Hegemonic Mestizaje, and Ambivalent Indigeneity in Proto-Chicana/o Autobiographical Discourse, 1858–2008." *MELUS* 38.1 (2013): 30–49.

Ontiveros, Randy J. *In the Spirit of a New People: The Cultural Politics of the Chicano Movement*. New York: New York University Press, 2013.

Otero Warren, Adelina. *Old Spain in Our Southwest*. New York: Harcourt, Brace, 1936.

———. "The Wind in the Mountains." Ms. Box 7, Folder 74. Family Professional Papers: Adelina (Nina) Otero Warren. Bergere Family Papers, 1829–1967. New Mexico State Archive. Santa Fe, New Mexico.

Paredes, Américo. *George Washington Gómez: A Mexicotexan Novel*. Houston: Arte Público Press, 1990.

———. *With His Pistol in His Hand: A Border Ballad and Its Hero*. Austin: University of Texas Press, 1970.

Paredes, Raymund A. "The Evolution of Chicano Literature." *MELUS* 5.2 (1978): 71–110.

———. "Mexican American Authors and the American Dream." *MELUS* 8.4 (1981): 71–80.

Parra, Andrea. Letter. *PMLA* 114 (1999): 1100.

Parras, Bryan, and Liana Lopez. Personal interview. 22 July 2015.

Paz, Octavio. *The Labyrinth of Solitude*. Trans. Lysander Kemp. New York: Grove Press, 1961.

Peña, Devon G. "The Scope of Latino/a Environmental Studies." *Latino Studies* 12.3 (2003): 47–78.

———. *Chicano Culture, Ecology, Politics: Subversive Kin*. Tucson: University of Arizona Press, 1998.

———. *Mexican Americans and the Environment: Tierra y Vida*. Tucson: University of Arizona Press, 2005.

Perez, Domino Renee. "The Politics of Taking: La Llorona in the Cultural Mainstream." *Journal of Popular Culture* 45.1 (2012): 153–71.

Phippen, J. Weston. "How One Law Banning Ethnic Studies Led to Its Rise." *The Atlantic* 19 July 2015. http://www.theatlantic.com/education/archive/2015/07/how-one-law-banning-ethnic-studies-led-to-rise/398885/.

Platt, Kamala. "Ecocritical Chicana Literature: Ana Castillo's 'Virtual Realism.'" *ISLE: Interdisciplinary Studies in Literature and Environment* 3.1 (1996): 67–96.

Porta-Gandara, Miguel Angel, Eduardo Rubio, and Jose Luis Fernandez. "Economic Feasibility of Passive Ambient Comfort in Baja California Dwellings." *Building and Environment* 37 (2002): 1001.

Powell, John Wesley. *The Exploration of the Colorado River and Its Canyons.* 1874. New York: Penguin Classics, 2003.

Preston, William. "Serpent in the Garden: Environmental Change in Colonial California." *Contested Eden: California Before the Gold Rush.* Eds. Ramón A. Gutiérrez and Richard Orsi. Berkeley: University of California Press, 1998.

Pulido, Laura. *Environmentalism and Economic Justice: Two Chicano Struggles in the Southwest.* Tucson: University of Arizona Press, 1996.

Raup, H. F. "Transformation of Southern California to a Cultivated Land." *Annals of the Association of American Geographers* 49.3.2 (1959): 58–78.

Ray, Sarah Jaquette. *The Ecological Other: Environmental Exclusion in American Culture.* Tucson: University of Arizona Press, 2013.

Rebolledo, Tey Diana. "Las escritoras: Romances and Realities." *Pasó por aquí: Critical Essays on the New Mexico Literary Tradition, 1542–1988.* Ed. Erlinda Gonzales-Berry. Albuquerque: University of New Mexico Press, 1989. 199–214.

———. Introduction. *We Fed Them Cactus.* By Fabiola Cabeza de Baca. 1954. Albuquerque: University of New Mexico Press, 1994. xiii–xxxii.

———. *Women Singing in the Snow: A Cultural Analysis of Chicana Literature.* Tucson: University of Arizona Press, 1995.

REI Foundation. *The REI Foundation Awards $410,000 to Connect Younger and More Diverse Groups to Nature.* 2014. 21 April 2014.

Reyna, Sergio. Introduction. *The Woman Who Lost Her Soul and Other Stories.* By Jovita González. Houston: Arte Público Press, 2000. x–xxvii.

Rivera, John-Michael. *The Emergence of Mexican America: Recovering Stories of Mexican Peoplehood in U.S. Culture.* New York: New York University Press, 2006.

Rivera, Raquel Z., narr. "Planet New Mexico." *LatinoUSA.* National Public Radio. KERA, Dallas. 5 Oct 2012.

Rivera, Tomás. *And the Earth Did Not Devour Him.* 1971. Trans. Evangelina Vigil-Piñón. Houston: Arte Público Press, 1995.

Rosales, F. Arturo. *Chicano!: The History of the Mexican American Civil Rights Movement.* Houston: Arte Público Press, 1996.

Ruffin, Kimberly N. *Black on Earth: African American Ecoliterary Traditions.* Athens: University of Georgia Press, 2010.

Ruiz de Burton, María Amparo. *The Squatter and the Don.* 1885. Eds. Rosaura Sánchez and Beatrice Pita. Houston: Arte Público Press, 1992.

———. *Who Would Have Thought It?* 1872. Eds. Rosaura Sánchez and Beatrice Pita. Houston: Arte Público Press, 1995.

Runte, Alfred. *National Parks: The American Experience*. Lincoln: University of Nebraska Press, 1997.

Saldaña-Portillo, María Josefina. *The Revolutionary Imagination in the Americas and the Age of Development*. Durham, NC: Duke University Press, 2003.

Sánchez, Rosaura, and Beatrice Pita. "María Amparo Ruiz de Burton and the Power of Her Pen." *Latina Legacies: Identity, Biography, and Community*. Eds. Vicki L. Ruiz and Virginia Sánchez Korrol. Oxford: Oxford University Press, 2005. 72–83.

"La Serpientita." *El Grito del Norte* 1.3 (1968): 8.

Shea, Anne. "'Don't Let Them Make You Feel You Did a Crime': Immigration Law, Labor Rights, and Farmworker Testimony." *MELUS* 28.1 (2003): 123–44.

Solis, Hilda. Foreword. *Hidden Danger: Environmental Health Threats in the Latino Community*. By Adrianna Quintero-Somaini et al. Washington, DC: National Resources Defense Council, 2004. iv.

Stanford University, Resources for the Future, and *New York Times*. *2015 National Global Warming Study*. 2015. Resources for the Future. 5 May 2015.

Stegner, Wallace. "The Legacy of Aldo Leopold." *Companion to* A Sand County Almanac*: Interpretive and Critical Essays*. Ed. J. Baird Callicott. Madison: University of Wisconsin Press, 1987.

Stephenson, Wen. "Looking for 'Climate Justice' on Houston's East Side." *The Nation* 25 July 2013. Web. 20 June 2015.

Thoreau, Henry David. *Walden*. 1854. Ed. Bill McKibben. Boston: Beacon Press, 1997.

Tijerina, Reies Lopez. *They Called Me "King Tiger": My Struggle for the Land and Our Rights*. Trans. José Angel Gutierrez. Houston: Arte Público Press, 2001.

Tonn, Horst. "*Bless Me, Ultima*: A Fictional Response to Times of Transition." *Aztlán* 18.1 (1989): 59–68.

Tulchin, Ben, and Corey O'Neil. *New California Statewide Survey Shows Conservation is a Key Latino Value*. 2012. California League of Conservation Voters. 8 May 2015.

Ulibarrí, Sabine. "Man Without a Name." *Tierra Amarilla: Stories of New Mexico/Cuentos de Nuevo Mexico*. 1964. Trans. Thelma Campbell Nason. Albuquerque: University of New Mexico Press, 1993. 2–81.

Urrea, Luis Alberto. *The Devil's Highway: A True Story*. New York: Little, Brown and Company, 2004.

U.S. General Accounting Office. *Treaty of Guadalupe Hidalgo: Findings and Possible Options Regarding Longstanding Community Land Grant Claims in New Mexico.* 2004. http://www.gao.gov/products/GAO-04-59.

Valdez, Luis. *The Shrunken Head of Pancho Villa. Necessary Theater: Six Plays About the Chicano Experience.* Ed. Jorge Huerta. Houston: Arte Público Press, 1989. 142–217.

Vasquez, Enriqueta. *Enriqueta Vasquez and the Chicano Movement: Writings from El Grito del Norte.* Eds. Dionne Espinoza and Lorena Oropeza. Houston: Arte Público Press, 2006.

Viramontes, Helena María. *Under the Feet of Jesus.* New York: Plume, 1995.

Wald, Sarah D. "Visible Farmers / Invisible Workers: Locating Immigrant Labor in Food Studies." *Food, Culture, Society* 14.4 (December 2011): 567–86.

Whaley, Charlotte. *Nina Otero-Warren of Santa Fe.* Albuquerque: University of New Mexico Press, 1994.

Wilbur-Cruce, Eva Antonia. *A Beautiful, Cruel Country.* Tucson: University of Arizona Press, 1990.

Worster, Donald. "John Muir and the Modern Passion for Nature." *Environmental History* 10.1 (2005): 8–19.

Yarbro-Bejarano, Yvonne. *The Wounded Heart: Writing on Cherríe Moraga.* Austin: University of Texas Press, 2001.

Ybarra, Patricia. "The Revolution Fails Here: Cherríe Moraga's *The Hungry Woman* as a Mexican Medea." *Aztlán: A Journal of Chicano Studies* 33.1 (2008): 63–88.

INDEX